# Advances in Science, Technology & Innovation

## IEREK Interdisciplinary Series for Sustainable Development

**Advances in Science, Technology & Innovation (ASTI)** is a series of peer-reviewed books based on the best studies on emerging research that redefines existing disciplinary boundaries in science, technology and innovation (STI) in order to develop integrated concepts for sustainable development. The series is mainly based on the best research papers from various IEREK and other international conferences, and is intended to promote the creation and development of viable solutions for a sustainable future and a positive societal transformation with the help of integrated and innovative science-based approaches. Offering interdisciplinary coverage, the series presents innovative approaches and highlights how they can best support both the economic and sustainable development for the welfare of all societies. In particular, the series includes conceptual and empirical contributions from different interrelated fields of science, technology and innovation that focus on providing practical solutions to ensure food, water and energy security. It also presents new case studies offering concrete examples of how to resolve sustainable urbanization and environmental issues. The series is addressed to professionals in research and teaching, consultancies and industry, and government and international organizations. Published in collaboration with IEREK, the ASTI series will acquaint readers with essential new studies in STI for sustainable development.

More information about this series at http://www.springer.com/series/15883

Khai Ern Lee
Editor

# Concepts and Approaches for Sustainability Management

*Editor*
Khai Ern Lee
Institute for Environment and Development
(LESTARI)
Universiti Kebangsaan Malaysia
Bangi, Malaysia

ISSN 2522-8714 ISSN 2522-8722 (electronic)
Advances in Science, Technology & Innovation
IEREK Interdisciplinary Series for Sustainable Development
ISBN 978-3-030-34570-9 ISBN 978-3-030-34568-6 (eBook)
https://doi.org/10.1007/978-3-030-34568-6



This Springer imprint is published by the registered company Springer Nature Switzerland AG
The registered company address is: Gewerbestrasse 11, 6330 Cham, Switzerland

# Preface

With the introduction of the 2030 Agenda for Sustainable Development by the United Nations General Assembly in September 25, 2015, UN agencies, member states, and stakeholders have begun to focus on the adoption and implementation of these strategies in realization of 17 Sustainable Development Goals. To work toward sustainability, strategic measures to encourage stakeholders to contribute to the goals of the 2030 agenda are needed.

In recognition of these efforts, this book is produced to compile research concepts and approaches for the area of sustainability management of industry, technology development, community, education, and environment. The objective of this book is to deliberate concepts and approaches of sustainability management taking place in Malaysia whereby case studies will be revealed to provide way forward of sustainability management toward achieving sustainable development.

The insights provided can be applied to advanced and developing countries by sustainable development practitioners, encompassing government agencies, academia, industries, NGOs, and community, who would like to adopt the concept of approach of sustainability into their area of management.

Bangi, Malaysia Khai Ern Lee

**Acknowledgements** The editor and authors would like to acknowledge the research grants provided by the Ministry of Higher Education and the National University of Malaysia (UKM) through Fundamental Research Grant Scheme (FRGS/2/2013/SSI12/UKM/02/1; FRGS/1/2019/WAB05/UKM/02/2), *Geran Galakan Penyelidikan Muda* (GGPM-2013-084), *Geran Universiti Penyelidikan* (GUP-2014-034 and GUP-2017-016), *Top-Down Research Grant* (TD-2014-015), *Dana Impak Perdana* (DIP-2015-008), *Ganjaran Penerbitan* (GP-K020155; GP-2019-K020155), and *Dana Industri* (XX-2019-006).

# Contents

# Understanding Public Benefit and Risk Perceptions Through Psychological and Sociological Aspects for Sustainable Nanotechnology Development in Malaysia

Nur Aizat Kamarulzaman, Khai Ern Lee, and Kim Shyong Siow

**Abstract**
Nanotechnology has opened a new realm to science and technology whereby it has been developed and used in various applications with potentials to facilitate sustainable development. The applications of nanotechnology are beneficial in improving public health, enhancing the functionality and endurance of consumer products, and potentially preserving the environment; however, uncertain risks associated with nanomaterials need to be understood for a good governance of nanotechnology to ensure sustainable development of nanotechnology. Hence, public perceptions of nanotechnology are instrumental for good governance of nanotechnology to determine the acceptance and rejection of the public toward nanotechnology. In this chapter, the public benefit and risk perceptions of nanotechnology are deliberated based on a case study in Klang Valley, Malaysia, through psychological and sociological aspects. Psychologically, knowledge is not a factor affecting the benefit and risk perceptions. However, the public perceives nanotechnology to be more beneficial than risky. Public attitudes are positive for nanotechnology, giving people benefit perception and reducing risk perception of nanotechnology. Trust in government, industry, and researchers increases the public benefit perception on nanotechnology as they are the driving force of nanotechnology development. The government as the regulator of nanotechnology development affects risk perception when public trust in government declines. Therefore, the government needs to play a role in getting public trust, thereby enhancing the public benefit perception on nanotechnology. Sociologically, i.e., culture, religious belief, and social aspect influence the public benefit perception but not risk perception on nanotechnology. Policy and religion emphasizing science and technology as an economic driver for the well-being bring the culture in receiving both scientific and technological developments in general and nanotechnology in particular. Correspondingly, continuous research of nanotechnology will result in the social implication by ensuring equal distribution of nanotechnology benefit, and at the same time, its risk will be effectively managed.

**Keywords**
Benefit • Risk • Perception • Nanotechnology • Sustainable development

## 1 Nanotechnology

Nanotechnology is defined as "technological research and development at atomic, molecular, or macromolecular levels, on a scale of 1–100 nm that provides a basic understanding of the material phenomenon at the scale of creating and using structures, devices, and systems that have the properties and novel functions because of its small size" (Roco 2011; Kamarulzaman et al. 2019). Small nanoscale has a wide surface to react effectively, compared to the same material but at the scale of hundreds of nano or microns. Nanomaterials can improve the previously unattainable electronic, optical, catalyst, and magnetic functions of a typical size within a range of hundreds of nanometers. The novel property of this material has function that can be processed into various forms, and physical chemistry of nanomaterials makes it widely used in manufacturing to develop more durable and high-performance products (Gleiche et al. 2006; West et al. 2016). The rapid development and widespread applications of nanotechnology make

N. A. Kamarulzaman · K. E. Lee (✉)
Institute for Environment and Development (LESTARI), Universiti Kebangsaan Malaysia, Bangi, 43600, Selangor, Malaysia
e-mail: khaiernlee@ukm.edu.my

K. S. Siow
Institute of Microengineering and Nanoelectronics (IMEN), Universiti Kebangsaan Malaysia, Bangi, 43600, Selangor, Malaysia



K. E. Lee (ed.), *Concepts and Approaches for Sustainability Management*, Advances in Science, Technology & Innovation, https://doi.org/10.1007/978-3-030-34568-6_1

it one of the catalysts for the Fourth Industrial Revolution (Maynard 2015).

Nanotechnology is developed as a capable technology for a variety of powerful applications that have never been thought to exist before which is also similar to the uncertain risks associated with nanotechnology. Many of the successes acquired by the new findings today would have a social impact on the public. Nanotechnology applications have changed the lives of people with more energy, telecommunication, medical, and engineering applications (Moussaouy 2018) whereby public is the one benefited and exposed to the risks of technology developed in a country. However, some risks are acceptable to the public when the risks are properly managed to bring benefits (Starr 1969). Nanotechnology has the potential to create value to sustainable development through public, economic, and industrial development if risks associated with nanotechnology can be controlled (Renn and Roco 2006). If the public engaged in the early stage of nanotechnology development, their perceptions will be taken into account in raising public awareness of nanotechnology and thus enabling policy-makers to develop nanotechnology in harmony with the needs of the public that will also ensure their well-being and safety (Rogers-Hayden and Pidgeon 2008). In this regard, this will encourage the advancement of technology that is in line with the progress of public thinking with the cooperation of stakeholders, such as the government, researchers, and industry players, to convey knowledge to the public. Hence, this partnership will create harmonious technology with the public. The development of nanotechnology not only provides the opportunities for public participation in voicing their will for a prosperous life and concern for the associated risks, but also opportunities for stakeholders to gain people's confidence by managing the risks posed nanotechnology effectively (Michelson and Rejeski 2006). As a result of collaboration between the public and stakeholders, it will create knowledge, skills and value to nanotechnology development for a sustainable future (Moussaouy 2018).

The benefits and risks of nanotechnology can be exposed to the public through nanotechnology applications. Nanotechnology has the potential for sustainable development by enhancing product functions, strengths, and prolonging product life span. However, there are concerns about uncertain risks associated with nanotechnology. The first thing to worry is the size of nanoparticles ranging from 1 to 100 nm, and this is the size that causes the nanoparticles to have different physical, chemical, and biological properties compared to the same but larger particles (Vishwakarma et al. 2010). These tiny nanoparticles size can penetrate into the human body and damage the cells and tissues. The second is the manufacturing and disposal of nanomaterials which may produce new pollutants that can be released into the environment through water and air (Roco 2003).

There are several nanotechnology applications that involve public directly, such as cosmetics, electrical appliances, foods, sports equipment, medicine, and household products (Kishimoto 2010; West et al. 2016). Table 1 shows the benefits and risks of nanotechnology applications.

Nanotechnology applications in cosmetic products and electrical appliances have been widely available in the market (Bennet-Woods 2008; Mamadou et al. 2012).

**Table 1** Benefits and risks of nanotechnology applications

| Nanotechnology applications | Nanomaterials | Benefits | Risks | References |
|---|---|---|---|---|
| Electrical appliances | Carbon nanotubes and quantum dots | Efficient use and storage of electricity | Exposes to hazard during the production of nanomaterials | Allsopp et al. (2007) |
| Medicine | Carbon nanotubes and boron nitride nanotubes | Diagnosis of illness using nanomaterials and drug delivery on cells | Toxic effect of nanomaterials to the cells | Raffa et al. (2010) |
| Detergent | Titanium dioxide, nanosilver | Dirt is easily removed and anti-bacterial | Endangers aquatic life | Mehic (2012) |
| Cosmetic | Nanoliposomes | Improves the absorption of cosmetic products into the skin | Absorption of nanomaterials into skin and respiration | Raj et al. (2012) |
| Food | Zinc oxide, carbon nanotubes, and titanium dioxide | Smart delivery of nutrition and nanoencapsulation of nutrients to plants | Health risk to consumers | Parisi et al. (2014) |
| Water treatment | Zeolite and titanium dioxide | Treats water by removing organic and inorganic compounds, microorganisms, and heavy metals | Nanomaterials can be new pollutants to the environment | De Luca and Ferrer (2017) |
| Sports equipment | Zinc oxide, carbon nanotubes, and titanium dioxide | Increase strength, lightness, and comfortness | Risk to users' skin | Harifi and Montazer (2017) |

Examples of cosmetic products containing nanomaterials are lipstick, face cream, toothpaste, and UV cream (Hristozov and Malsch 2009). Nanoliposome is used to improve the absorption of cosmetic products into the skin and thereby improve product efficiency (Raj et al. 2012). This widespread use concerns the consumers about the safety of nanomaterials contained in cosmetics that can seep into the skin and enter the respiratory tract when inhaled (Mu and Sprando 2014).

Electronically assembled hardware also contains nanomaterials, such as carbon nanotubes and quantum dots to promote efficient use and storage of electricity (Allsopp et al. 2007; Mensch and Umwelt 2014). However, consumers are at risk exposing to nanomaterials through air and skin contact, when they are working in the production of nanomaterials in a laboratory or manufacturing plant.

With the growth of world population, food and water security are the fundamental issues that need to be addressed to achieve sustainable development. Nanomaterials such as zinc oxide are applied in fertilizers to improve nutrient absorption by plants. Nanosensors are used to identify early disease on plants to prevent crop damage (Handford et al. 2014). Nanotechnology applications in food production have the capability to increase food production and avoid the shortage of food sources in the future (Parisi et al. 2014). However, the impacts of nanomaterials to consumers' health and the environment are of concerns of the public.

Nanotechnology is used to treat water including groundwater and wastewater by using carbon nanotubes, zeolites, and titanium dioxide which are capable in removing organic and inorganic compounds, microorganisms, and heavy metals (De Luca and Ferrer 2017). Magnetic nanoparticles that are widely used in treating water can be easily obtained from the nature; hence, this makes the cost of using nanomaterials in water treatment cheaper than conventional ones (Fromer and Diallo 2013; Sannino et al. 2017). Water treatment using nanotechnology is cost-effective in ensuring adequate sanitation and water supply for the world's population. However, nanomaterials could be a new pollutant to the environment through water drainage (Roco 2003).

Sports activities are part of a healthy lifestyle for the well-being of the public, while advanced medicine can safeguard the public health. Nanomaterials such as carbon nanotubes, zinc oxide, and titanium dioxide are used in sports equipment to increase strength, lightness, and comfort to consumers. Among other benefits being sports clothing designed with nanotechnology have advantages, such as waterproof, anti-bacteria, anti-odor, and UV protection. However, the risk on consumers' skin and the environment is still unknown (Harifi and Montazer 2017). In advanced medicine, carbon nanotubes and boron nitride nanotubes are used in nanodiagnosis through magnetic resonance imaging and therapy using nanomaterials to improve diagnosis and treatment of diseases (Raffa et al. 2010). However, the use of nanoparticles for the delivery of drugs to certain cells has the risk of harmful substances from the nanoparticles. Nanoparticles are likely to be released into the body and cause toxic effects to cells and tissues.

Nanoscale titanium oxide and silver are known for their anti-bacterial and hydrophobic properties that can be applied in detergents, such as dishwashing soap and detergent soap (Mehic 2012). The detergent will be more effective when the dirt is easily removed, and the cleaned surface becomes waterproof. Thus, the surface will take longer time to become dirty again (Gleiche et al. 2006). Compared to larger-sized silver, nanosized silver has a higher free radical rate and when nanosized silver flows into the body of water, it can endanger the living (Mehic 2012).

The application of nanotechnology is beneficial to the public for the well-being of life and improves the public health which is one of the important elements for sustainable development. However, uncertain risks associated with nanomaterials are alarming and require a good governance of nanotechnology to ensure the safety of the public. Hence, public perceptions of nanotechnology are instrumental for good governance of nanotechnology to further understand the needs of the public toward nanotechnology. This understanding is essential for nanotechnology to be developed without leaving behind the public that can lead to rejection and thus inhibit sustainable development of nanotechnology.

## 2 Public Perception Toward Nanotechnology

The appropriate use of nanotechnology can address environmental limitations to meet the needs of population growth. From time to time, nanotechnology developed will undergo several modifications to meet the acceptance of the public. This modification will continue to take place so that nanotechnology can be fully accepted by the public, and the technology is said to have reached its stable and sustainable use by the public (Saidi 2018). The benefit and risk of nanotechnology will be evaluated by the public, whereas the conflict in the acceptance and rejection of nanotechnology will depend on differences in interpretation and controversial cases that are associated with nanotechnology.

Public perception is defined as a social phenomenon of how public sees risks and benefits in current situations based on facts or fictions of current knowledge, culture, and/or media. Public processes risk in their minds as a concept to deal with uncertainty and danger in life (Sjöberg et al. 2004). Conversely, benefit perception is built up when they believe it will get something positive based on a specific action (Leung 2007). The public and those who are experts on nanotechnology have a very different way of looking at the risks. From experts' point of view, risk is seen as an annual death rate, while the public sees it as a level of danger and accidents.

The human mind builds uncertainty because the lack of complete knowledge put individuals in an unconfident position with fear or suspicious. Individuals will choose to act in a safe condition and less risky to them. However, not all risks are acceptable to the public. Some risks are acceptable if it brings benefits (Starr 1969), and this type of risk is called voluntary risk. Mathematically, risk is defined as the magnitude of the loss multiplied by the probability of occurrence. Risk is viewed from various angles including risk factors, time risks, and those that will be affected by certain risks. The subjective assessment of the probability of occurrence creates the individual's benefit and risk perceptions. The benefit and risk perceptions are indeed a debate whether the perception should be rationalized by individuals who are experts in a particular field or acceptable from the irrational opinion of the public opinion (Fischhoff et al. 1983). Individual's perceptions of nanotechnology are not only influenced by their level of knowledge, but their perceptions are also influenced by social, cultural, and ideological (Sjöberg et al. 2004).

The heuristic cognitive strategies are most commonly used by individuals to make choices and assessments quickly (Gilovich et al. 2002; Pieper 1989). Individuals will use heuristics by customizing existing information to assist them in making decisions. When a person receives information stating low risk is associated with a technology, the person will evaluate the technology as safe and vice versa (Slovic et al. 2007). Individuals will not only evaluate based on the information provided, they will also evaluate based on what they feel, whether they like or dislike a particular technology (Finucane et al. 2000). However, decisions made using heuristic methods can be biased and will contribute to errors (Stanovich and West 2000). However, today we need to make a quick decision based on what has been presented to us (Gilovich et al. 2002); hence, different levels of intelligence, worldview, and thinking will result in different decisions.

In the early stage of understanding the perceptions of nanotechnology, people with a scientific background feel that nanotechnology is beneficial, but some worry about the inequality of future benefits (Bainbridge 2002). Concerns about the risks in nanotechnology development have increased in 2003 when research on nanotechnology impacts on the public, such as new pollutants produced in nanoscale began to be published (Roco 2003). Concerns about misuse of nanotechnology produce more destructive weapons than weapons that do not use nanotechnology (Phoenix and Treder 2003). The public concerns over the same risk of nanotechnology (Cobb 2005; Cobb and Macoubrie 2004; Cormick 2009; Macoubrie 2006) whereby not only weapons are the main concerns of the public, but also devices that can affect the privacy of the public, such as mini-surveillance devices that can be placed in clothing and stuff (Cobb and Macoubrie 2004; Cormick 2009). In addition, "fear of something unknown," "acts contrary to nature's processes," and "environmental degradation" are among the public's concerns about nanotechnology (Cormick 2009).

Meanwhile, nanotechnology has been proven to give solutions to many problems, such as early detection and disease treatment, optimizing the use of non-renewable resources, effective pollution recovery, and many other benefits (Kharat et al. 2017). Researches show that people respond positively to nanotechnology compared with worry (Bainbridge 2002; Binder et al. 2012; Bostrom and Löfstedt 2010; Burri and Bellucci 2008; Cobb and Macoubrie 2004; Cormick 2009; Dijkstra and Critchley 2014; Zhang et al. 2015) though they are not knowledgeable about nanotechnology. Based on the limited knowledge, the public can still make decision based on what is provided to them by the media and use other social aspects to assess the benefits and risks of new technologies (Lee et al. 2005; Schütz and Wiedemann 2008). The individual's ability to assess nanotechnology whether is beneficial or risky will determine the acceptance or rejection of the public. The public perceptions are important to enable the continued development of nanotechnology for sustainable development; hence, the factors that influence the public perceptions will guide the government, researchers, and industry to understand the needs of the public and develop nanotechnology sustainably.

## 3 Psychological and Sociological Aspects for Public Benefit and Risk Perceptions

The heuristic method introduced by Simon (1977) has suggested a cognitive strategy to make a decision easily under uncertain circumstances. When a person makes a decision, the individual cannot avoid making decisions that are influenced by personal, socioeconomic, and political views, cultures, and so on that become part of a person's life (Pieper 1989). People have different perceptions when exposed to the same information but presented in different ways (Tversky and Kahneman 1981). Wildavsky (1987) argues that one does not have to work hard and become a politician to give opinions on politics, but they need to know only some of the information they have and they already can give opinions (Wildavsky 1987).

While cultural theory explains the tendency of individuals to make choices whether beneficial or risky for a dangerous thing or activity is dependent on the culture practiced (Kahan et al. 2009). Kahan et al. (2009) also shows that individuals choose information that is relevant to the cultural views as well as political inclination despite being exposed to the same information. The concept of cognitive psychology and cultural theory is the basis that adapts two approaches, namely psychology and sociology, from Renn and Swaton (1984) to better understand public perceptions of nanotechnology. Psychological aspects focus on cognitive psychology involving one's cognitive ability to assess risks and benefits when making decisions based on knowledge, individual attitudes, and trusts to stakeholders managing nanotechnology (Renn and Swaton 1984). While sociological aspects are the decisions made by individuals who are influenced by social groups represented by them (Renn and Swaton 1984) whereby this approach encompasses culture and religious beliefs they practice. In addition, there are intervening variables (moderators) which also influence the public perceptions of nanotechnology psychologically and sociologically, such as media coverage, technology development, and economic status, different nanotechnology applications, as well as benefits and risk information (Petersen et al. 2007; René Zimmer et al. 2010; Schütz and Wiedemann 2008; Siegrist 2010).

## 3.1 Psychological Aspects

Psychological aspects refer to cognitive psychology involving the receipt of information, retention of information, and retrieval of necessary information. However, information may be interpreted differently depending on respective individual whereby individual with complete information and knowledge will make more objective decision than an underprivileged individual. This situation differs from individual who does not have enough information to make decisions that are biased based on their preference (Finucane et al. 2000). However, the biased decisions are still accepted as one of the ways in thinking whereby a person's failure to express feelings because impairment of the brain causing the individual not able to make decision and socialize (Damasio et al. 1990).

Experts like scientists and researchers who are knowledgeable about nanotechnology show different perceptions about nanotechnology compared to the civil society (Cormick 2009; Siegrist et al. 2007a, b). Both of these groups of experts agree that the benefits of nanotechnology exceed the risks even though people tend to see nanotechnology has more risks than experts do. Those with complete knowledge of nanotechnology will find the benefits exceeding the risks and are willing to accept the technology (Binder et al. 2012; Brossard et al. 2009; Retzbach et al. 2011; Scheufele and Lewenstein 2005). Based on experts' knowledge, they are less concerned about the risks of nanotechnology as the risks of nanotechnology have no direct impact on the civil society (Siegrist et al. 2007a, b). Complete knowledge will put individuals in a confident position without fear. The difference in the perceptions of benefits and risks between experts and the public is also due to experts' experience which is not available to the public. Their experience, knowledge, and expertise in handling nanotechnology enable them to see the development of nanotechnology in a controlled manner resulting in their risk perception lower than the public (Renn and Swaton 1984). However, experts are concerned with "new pollution" and "new disease" that may arise due to nanotechnology in the future (Cormick 2009).

Limited knowledge of the public about nanotechnology does not refrain them from being positive with the development of technology when they have confidence in those managing nanotechnology. The trust given by the public to the government, researchers, and industry makes it easier for them to make decisions based on information received about nanotechnology (Chen et al. 2013). Public policy can also be easily formed with the trust given by the public (Pidgeon et al. 2009). According to Siegrist et al. (2007a, b), experts have more faith in the government in handling nanotechnology and protecting the public from risks. While the public depends on how the government, researchers, and industry manage the risks inherent in nanotechnology, people also tend to be hesitant and afraid of the goals of nanotechnology and the effects of nanotechnology they will receive. Citizens who trust government agencies like the Food and Drug Administration (FDA) will continue to believe the government will prioritize the public's interest by providing information on nanotechnology in labeling and possibly establishing a mandatory labeling in the future (Brown and Kuzma 2013). The public trusts are important for policy-makers to enable them to manage the public's doubts and concerns and move on to develop nanotechnology in the direction of convincing the public.

The attitude functions as a heuristic signal to individuals when one recalls their behavior in the past and influences their judgments and decisions in the present situation (Bem and McConnell 1970; Pratkanis 1988; Ross et al. 1981). The information received is defined differently according to the individual's background. A person will decide whether to like the information they receive based on their memory of the information. Heuristic attitudes are defined as "evaluative relationships in which one uses an object as a strategy to solve the problem by determining whether the object is in a preferred category (such as a strategy of liking, approaching and protecting) or being in an unwanted category (such as a strategy of hate, avoid and harm)" (Pratkanis 1988). The positive or negative attitudes of the public toward nanotechnology differ depending on the benefits and risks seen by individuals (Besley 2010; Chen et al. 2013). People who have a positive attitude toward science and technology and have never encountered a bad controversial experience involving science and technology in their country will see nanotechnology as beneficial, thereby having a positive attitude toward nanotechnology (Zhang et al. 2015). In contrast, people who faced controversial issues with science and technology will see nanotechnology as risky and likely to reject nanotechnology (Bennet-Woods 2008).

## 3.2 Sociological Aspects

Apart from the knowledge gap between experts and the public, the beliefs held by the public toward those managing nanotechnology, heuristic attitudes that produce risk and benefit perceptions, and personal values also operate as a perception filter for people to understand the emergence of new technologies. Individuals from diverse backgrounds will interpret the same information differently depending on the value and tendency they hold. Sociological aspects refer to the perceptions of the public that are created and influenced by social groups represented by a particular individual and also based on the cultures and the religions practiced.

An individual's way of life is a combination of cultural values (shared values and beliefs) and social relationships (human relationships) (Douglas 1978). Culture is defined as a lifestyle that contains values and beliefs taught by older generations and then inherited by the younger generation (Oltedal et al. 2004). In addition, one's views are also shaped by social groups, such as organizations and peers who are part of the individual (Tansey and O'Riordan 1999). The public learns and understands their culture in deciding whether something is beneficial or risky and choose something according to their cultural values (Mamadouh 1999). Solid support for nanotechnology can be observed in China whereby Zhang et al. (2015) shows that Chinese have high expectations on nanotechnology to improve their standard of living and enable them to compete globally. It has become a culture for the Chinese society to support technology since the establishment of the Open-Door Policy in 1978 which involved science and technology as the prime mover for economic and industrial development.

In the social context for social interactions and social relationships in social groups (a group of individuals with similar aims and collective unity), public perceptions of nanotechnology not only focus on nanotechnology toxicological risks but extend to the benefits and risks of nanotechnology in manufacturing and production, distribution, use and disposal of products that will only be experienced by certain social groups and not other groups (Conti et al. 2011). This leads to public perceptions of nanotechnology that involve the distribution of nanotechnology equally to all different societies and gender groups across various social groups. The inequality of the distribution of the benefits and risks of nanotechnology to social groups in society may lead to a handful of social groups tend to be exposed to risks rather than benefits of nanotechnology. Meanwhile, affordable social groups have access to nanotechnology benefits without being exposed to risks.

Different social groups will have different perceptions depending on how the technology benefits them and how risks are handled. Furthermore, social groups with knowledge of nanotechnology also have different interpretation of benefits and risks based on their technical skills in handling the technology (Saidi 2018). In this situation, social beliefs are best suited for the public when there is a lack of capacity, knowledge, interests, and resources. Social beliefs are the willingness to rely on those responsible for decision making and taking action relating to technology, environmental, medical or health management, and public safety management (Siegrist et al. 2000). Therefore, public acceptance of new technologies depends on information provided and social trust guides to the government, researchers, and industries (Cobb and Macoubrie 2004; Currall 2009; Siegrist et al. 2000).

Religion is part of a value system that affects individuals to understand new facts that are received including science and technology. Public acceptance of technology varies based on faithfulness of religion. People with a strong religious faith are morally disproportionate with nanotechnology as compared to those who have less faith in religion. The benefit perception toward nanotechnology is found among the less religious public and living in a secular society. Technology is a moral issue that gets the attention of the public holding religious beliefs because for them technology interferes the natural processes of nature. This situation is considered to be risky and morally unacceptable (Scheufele et al. 2009) because the public that hold strong faith also do not support funding for research and development of nanotechnology (Brossard et al. 2009) as they believe the technology interferes natural processes and associate with the term of "play God."

## 3.3 Other Factors

With limited knowledge of nanotechnology, heuristic is the common method used by people to make decisions. Television, radio, Internet, and newspapers are easily accessible to the public and are the main sources of information for the public (René Zimmer et al. 2010). Exposure to media coverage about nanotechnology will increase the public's understanding of the technology. Typically, mass media often reports something new with negative tone (Metag and Marcinkowski 2014) whereby this negative information that is over-showing about a technology will cause the public to become phobia toward the technology and causes the rejection of the technology.

Public perception depends on how the media display information about nanotechnology (Cacciatore et al. 2011;

Lee et al. 2005). Based on the information presented, the public will assess and decide whether nanotechnology is beneficial or risky. Media coverage can be in a different tone whether reporting on the benefits of nanotechnology exceeds risks by choosing only certain risks to highlight or report on nanotechnology risks exceeds its benefits (Gorss 2008). In addition to news tones that may affect public perceptions, scientific information should also be covered so that the public can develop benefit or risk perceptions objectively (René Zimmer et al. 2010).

The media that deals with the benefits of nanotechnology which covers its impact on social, economic, ethical, and related risks will provide complete input to the public. Subsequently, their perceptions will be formed based on a better understanding of the current development of nanotechnology (Tyshenko 2014). However, the public have the freedom to choose information they want to know and have their own interpretation (Lemanczyk 2014). Media coverage gives the public an idea of the development of new technologies that can influence the perceptions of the public at an early stage; however, as the technological development goes on, public perceptions may change accordingly (Nisbet and Huge 2007).

Public perceptions evolve as technology progresses. When technology is developing, more information about technology will reach the public and give them more understanding of the technology (Kahan et al. 2009). People have different perceptions and reactions to technologies they already know and understand. Low risk perceptions are found among societies in developed countries that adopt technology. Along with the development of technology, environmental destruction is something that worries a society that is not familiar with certain technologies as they view technology is threatening the environment and causing destruction. Compared to people in a country that is familiar with technology, environmental destruction is not a major cause of technological development (Lima et al. 2005). Fear of technology will decrease when people become more familiar with the technology and find benefits exceeding risks. In countries which economies are driven by technology, the public have benefit perception over their risk perception (Liu et al. 2009; Zhang et al. 2015). Public support for nanotechnology depends on their expectations of nanotechnology contributions to socioeconomic well-being (Könninger et al. 2010).

Although nanotechnology is generally accepted by the public, the public sees the benefits and risks of nanotechnology differently, and these differences depend on specific applications (Siegrist et al. 2008). The public accepts nanotechnology applications when they are useful, beneficial, and essential to them (Gupta et al. 2012). While nanomaterials in food can further enhance nutrient absorption into the body, nano-containing foods are of the most worried nanotechnology applications and most not accepted by the public (Kishimoto 2010; Siegrist et al. 2007a, b; Siegrist et al. 2008). The public believes nanomaterials in food are hazardous and can be harmful to health in the long run (Giles et al. 2015; Siegrist et al. 2007a, b). Electrical appliances containing nanomaterials and nanotechnology applications in medicine, such as channeling drugs to specific target cells, are accepted nanotechnology applications as they are beneficial to the environment and universal health (Gupta et al. 2012). Comparing to food packaging containing nanomaterials, the public finds that food packaging containing nanomaterials is more beneficial than food containing nanomaterials (Siegrist et al. 2007a, b).

The benefits and risks' information about nanotechnology products through labels are used to inform the public about the content of nanotechnology products and increase their awareness of products containing nanomaterials. Providing nanotechnology risks' information to the public increases the public awareness as the negative information has more impact compared to benefits information (Cobb 2005). The provided labels act as information tools to the public that help them to make decisions and select products in guided way (Chuah et al. 2018). The information provided will give the public an overview of the benefits and risks inherent in nanoproducts. Product labeling with "contains synthetic nanoparticles" label provides product risk information, hence increases consumer risk perception rather than non-labeled nanoproducts (Cobb 2005; Siegrist and Keller 2011). Labels help the public to make objective decisions without bias even though the information on labels is limited. Rational decisions made based on limited information are better than just relying on intuition alone without any fact of support (Renn and Swaton 1984). However, without any knowledge of nanotechnology and if users do not read labels on the product, nanotechnology product labeling can be ineffective as a medium to convey information to the public (Brown and Kuzma 2013).

## 3.4 Concept

The development of nanotechnology has shifted from research in laboratories to commercialization of products in the market. In 2020, nanotechnology is expected to increase up to USD 75.8 billion from nanoapplications in the fields of electronics, energy, cosmetics, medicine, defense, food, and agriculture (Sheila 2017). Nanotechnology is also referred as an industrial revolution that is potentially for sustainable development (Gaskell et al. 2005; Leinonen and Kivisaari 2010; Wiek et al. 2012). Public perception of nanotechnology on health, safety, and environment is an important factor to be addressed to ensure the responsible development of nanotechnology and meet the needs of today's generation without compromising the needs of future generations.

Technology transfer between industries and researchers not only produces products that use competitive nanotechnology in the market but also examines their impact on human health and the environment throughout the lifecycle of nanotechnology products (Musazzi et al. 2017). The increasingly advanced nanotechnology knowledge among experts and the rapid development of nanotechnology in the industry, however, have led the public to lag behind current technological developments. Previous studies show that educated individuals about technology will continue to increase their knowledge of nanotechnology, while less educated individuals will continue to lag behind (Corley and Scheufele 2010).

This gap of understanding and knowledge makes the use of new technologies, such as nanotechnology to be applied in the life of the public effectively becomes vague. Discussions between experts in the field of nanotechnology with civil society are the approaches taken to convey information to the public, while at the same time gaining the perspectives and views of the public on nanotechnology (Kass 2001). This will lead to good governance of nanotechnology that promotes public involvement to enable nanotechnology to be applied to the well-being of the public, thus leading to sustainable development. Good governance involves the process of making and implementing the decisions needed to develop nanotechnology properly. Good governance is defined as effective governance with specific characteristics, and governance performance needs to be assessed with accurate data (Rotberg 2014). There are eight characteristics of good governance, being (1) public involvement, (2) the rule of law with a fair and equitable legal system, (3) transparency in decision making and any action taken is in accordance with the law, (4) provide feedback which is effective at the right time, (5) consensus oriented of the parties involved, (6) the equality and inclusion of all parties involved, (7) effective and efficient in managing and using resources sustainably, and (8) accountability for each effect of the decision taken (UNESCAP 2009).

Good governance in this context is collective in managing nanotechnology at all levels of the organization to establish relationships not only among organizations but also with the public. Good governance of nanotechnology requires transdisciplinary knowledge and implementation between scientists and non-scientists (Hurni and Wiesmann 2014). The involvement of those with the skills of the public can reduce the knowledge gap between them (Roco et al. 2011), and decisions are made with social-oriented interests without prejudice (Rist et al. 2007). Procedures for decision making on risk-related issues require transparency in the administration and public involvement (Renn and Swaton 1984). Hence, the public benefit and risk perceptions toward nanotechnology give a preliminary view of the acceptance or rejection of the public which is the basis of good governance of nanotechnology so that the benefits of nanotechnology are distributed equally at all levels of society while the risks are well managed to ensure the safety of society.

The conceptual framework as shown in Fig. 1 provides an understanding of the two approaches, namely psychological and sociological approaches, that influence the public perception toward nanotechnology. The concept of

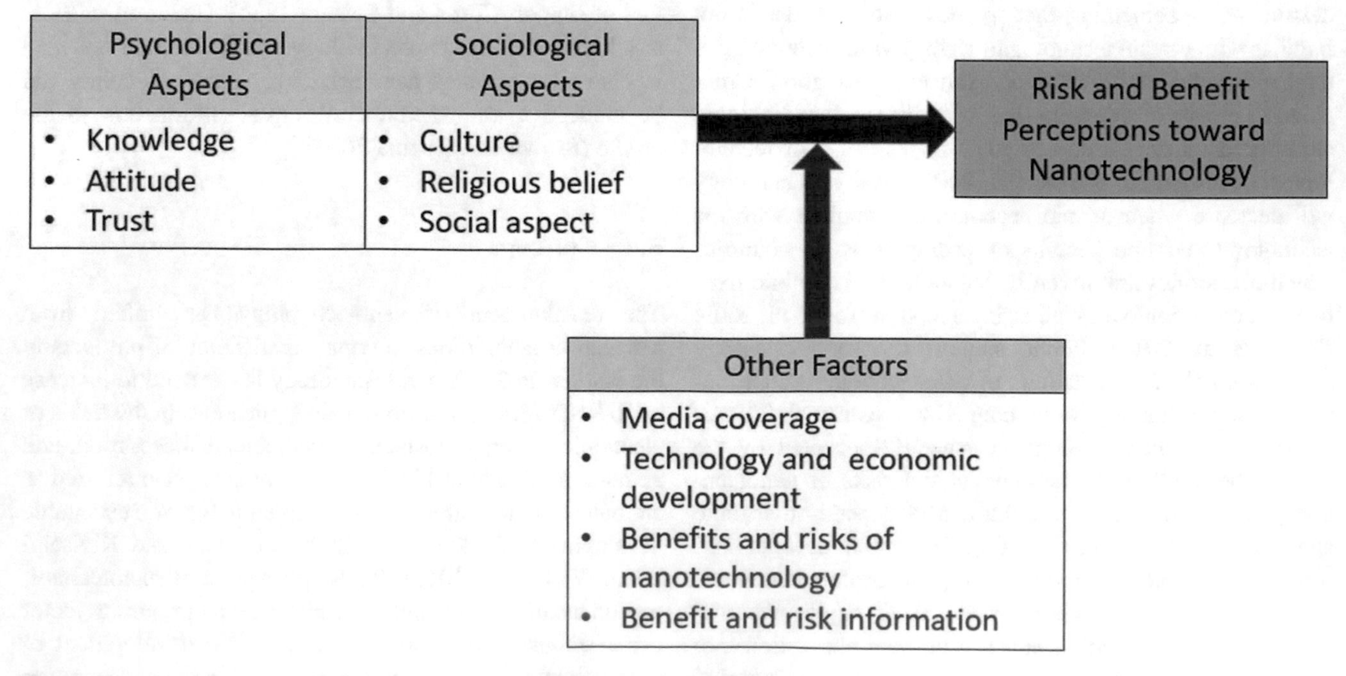


**Fig. 1** Conceptual framework for understanding psychological and sociological aspects that influence the risk and benefit perceptions toward nanotechnology (Kamarulzaman et al. 2018)

psychological and sociological approaches is adapted from Renn and Swaton (1984) that used these approaches to understand the perception of risk among society. Psychological aspects consist of knowledge, attitude, and trust aspects, while sociological aspects consist of culture, religious belief, and social aspects. Apart from these two approaches, there are also other factors affecting public perceptions known as moderators consisting media coverage, technological development and economic status, benefits and risks of nanotechnology applications as well as benefit and risk information. Understanding all these factors in influencing the public benefit and risk perceptions will be instrumental for sustainable nanotechnology development in Malaysia.

Following is the definition for perception and factors contributing to benefit and risk perceptions.

Public perception:

(i) Benefit perception—Benefit perception illustrates the mental process of representing and assimilating the possibility of beneficial events relating to certain objects or activities that may occur in the future (Renn and Swaton 1984).
(ii) Risk perception—Risk perception describes mental processes representing and assimilating the likelihood of adverse events associated with certain objects or activities that may occur in the future (Renn and Swaton 1984).

Psychological aspects:

(i) Knowledge—Knowing the definition, usefulness, and application of nanotechnology (Kishimoto 2010).
(ii) Attitude—The views and opinions of the public on nanotechnology are either beneficial or not including the application of nanotechnology into consumer products as well as the desire to purchase products that contain nanomaterials (Kishimoto 2010). Attitude is also a heuristic signal when one remembers the past behavior that will affect the decisions made today (Bem and McConnell 1970; Pratkanis 1988; Ross et al. 1981).
(iii) Trust—Trust refers to the government, industry, and researchers in developing nanotechnology toward meeting the needs and wants of the public (Kishimoto 2010).

Sociological aspects:

(i) Culture—Culture is defined as a lifestyle with shared values and beliefs inherited by older generations to young people who influence one's views and perceptions on something (Oltedal et al. 2004).
(ii) Religious belief—Religious belief is part of a value system that affects individuals in understanding new facts received including science and technology (Kim 2017).
(iii) Social aspect—Social groups, such as organizations and peers that influence one's views and perceptions on a subject including religious groups they represent (Tansey and O'Riordan 1999).

Other factors:

(i) Media coverage—Media coverage includes print and electronic media covering the benefits and risks of technology in general and nanotechnology in particular that affect the public's response to new technologies (Schenk et al. 2008).
(ii) Technological development and economic status—The development of a growing technology is capable of improving quality and economics that can facilitate the day-to-day affairs of the nation's development (Zhang et al. 2015)
(iii) Benefits and risks of nanotechnology—The benefits and risks of nanotechnology in applications that are used by the public (Kishimoto 2010).
(iv) Benefit and risk information—Benefit and risk information available on consumer products (Siegrist and Keller 2011).

## 4 Case Study: Klang Valley, Malaysia

The Ministry of Science, Technology and Innovation (MOSTI) had previously identified nanotechnology as one of the areas of research priorities and had spent more than 33.5 million USD on nanotechnology-related projects (Hashim et al. 2009). These investments were made to maintain Malaysia's market competitiveness in advanced materials, electronics, nutrition, cosmetics, medical designs, and various applications (Piccinno et al. 2012). In the early stages, nanotechnology was used to produce nanosized gold, silver, zinc, titanium, and black carbon, and now innovation has grown in nanotechnology covering nanotubes, graphene, and so forth (Lee et al. 2013).

The National Nanotechnology Initiative (NNI) Malaysia was launched in 2006 with a vision focusing on nanotechnology for the development of national science, technology, industry, and economy. NNI Malaysia was established to enhance Malaysia's economy, accelerate improvements, and enhance contributions to the public and the environment through the development of nanotechnology by gathering resources and knowledge among researchers, industry, and government. In 2007, a study funded by the Economic Planning Unit (EPU) released a report entitled "Identification of business and R&D opportunities in the applications of nanotechnology in Malaysia" suggesting nanoelectronics as one of the nanotechnology pioneering applications to be developed in Malaysia (Masrom 2012).

The National Nanotechnology Center (NNC) was set up in 2009 to foster nanotechnology activities under the auspices of NNI Malaysia. NNC's initiatives cover National Nanotechnology Statement, National Nanotechnology Center of Excellence, NanoMalaysia Limited, and NanoMalaysia Center in Iskandar, Malaysia. The National Nanotechnology Policy Statement states that "it is expected that nanotechnology will be a strategic growth engine for Malaysia, which will be achieved through a symbiotic national nanotechnology ecosystem that will ensure sustainable development." Subsequently, the Malaysian government launched the National Nanotechnology Statement in July 2010 to use nanotechnology as an engine that enables new economic growth and sustainable development that ensures the well-being of the public (Masrom 2012).

NanoMalaysia Limited was incorporated in 2011 as a limited company with a guarantee to act as a business entity in running nanotechnology commercialization activities such as managing NanoMalaysia Center and other NNC-approved infrastructure, commercialization of nanotechnology products, planning and coordination, research and development and international investment for nanotechnology, international networks and marketing of nanotechnology industry Malaysia, and products in global supply and value chains.

The efforts to enhance the development of nanotechnology in Malaysia have led to the establishment of the National Nanotechnology Center of Excellence in 2011 to support research and provide facilities and training. Several nanotechnology research centers were established at universities in Malaysia, such as Ibnu Sina Institute for Fundamental Science Studies (IIS) at Universiti Teknologi Malaysia, Institute of Microengineering and Nanotechnology (IMEN) at Universiti Kebangsaan Malaysia, Center of Innovative Nanostructures and Nanodevices (COINN) at Universiti Teknologi PETRONAS, Institute of Nano Electronics Engineering (INEE) at Universiti Malaysia Perlis, NEMS/MEMS Research Laboratory at MIMOS, Nano-Opto-Electronics Research Lab (NOR LAB) at Universiti Sains Malaysia, Institute of Advanced Materials (ITMA) at Universiti Putra Malaysia, and Nanotechnology and Catalysis Research Center (NANOCAT) at Universiti Malaya. Laboratory facilities set up at these research centers provide Malaysia an environment to develop toward a transdisciplinary technology country to achieve sustainable development as stated in the National Nanotechnology Statement (Masrom 2012).

Since nanotechnology has been identified as one of the new country's growth engines, the applications of nanotechnology in various fields help to improve the country's economy and thus drive sustainable development making it one of the technologies available for the Fourth Industrial Revolution (Schwab 2015). The Fourth Industrial Revolution will continue to grow rapidly, and Malaysia should follow such developments as it is envisaged in the Eleventh Malaysia Plan (Zainal Abidin 2018). Now, nanotechnology continues to advance from research to industry and is then commercialized in the form of products to consumers. The market value of nanotechnology commercial revenue is expected to increase by up to 10% in the years to come (Roco et al. 2011). This increase will certainly impact the country's economy and the daily lives of the people. However, local public still does not know about nanotechnology capabilities and is uncertain about the risks associated with nanotechnology (Chen et al. 2013; Cobb and Macoubrie 2004; Kishimoto 2010).

With the increased use of nanotechnology in manufacturing of products in Malaysia, the benefit and risk perceptions of society need to be understood to determine the acceptance of the society on nanotechnology. The increased understanding of science and technology that is being developed in Malaysia will encourage the public to participate in decision making. Positive attitude and public acceptance of new technologies such as nanotechnology are important in ensuring nanotechnology capabilities to continue to grow (Siegrist 2010). Public reactions to nanotechnology also involve policies and regulations designed to prioritize public interest in terms of safety, health, and environmental pollution prevention (Burri and Bellucci 2008; Cormick 2009; Siegrist 2010). Since Malaysia aspires to become a nanotechnology development hub and the Klang Valley is central to the development of nanotechnology, lack of knowledge and awareness on nanotechnology will lead to the deterioration of public involvement and trust to the government and inhibit the development of nanotechnology in Malaysia. Therefore, Klang Valley is taken as a case study to understand the perception of the society on nanotechnology which will be an important instrument for policy-makers, industry, and researchers to develop nanotechnology with good governance toward sustainable development that is compatible with the will of the public and thereby ensure the interests and well-being of society in Malaysia.

# 5 Public Perception Toward Nanotechnology Development

## 5.1 Perception Based on Demography

To understand psychological and sociological aspects in influencing public risk and benefit perceptions on nanotechnology, a stratified sampling survey was conducted on the public respondents in Klang Valley. This study tested on ten (10) different demographic categories, i.e., gender, age, marital status, race, residency area, religion, education, stakeholders, household income, and occupation. Table 2 shows only the significant demography that has effect on public benefit and risk perceptions. Race and level of education are shown to have significant influence on both public benefit and risk perceptions. This study on the other hand is focusing on the race, but ethnicity is not included. There are significant differences between the majority race (Malay) with other minorities (Chinese and Indian) in terms of benefit and risk perceptions as the Malay feels a lot more secured with nanotechnology. Highly educated individuals have more confidence in the government and researchers to protect them from the risks of nanotechnology, thus perceiving more benefit than risk in nanotechnology. This can be comparable in another study conducted by Cobb and Macoubrie (2004) where whites and educated individuals perceive nanotechnology more beneficial than risky.

Next, age and household income only influence public risk perception but not benefit perception. Younger respondents in this study have lower risk perception because youngsters are more optimistic about nanotechnology (Gaskell et al. 2005). George et al. (2014) discover that public over the age of 48 and under the age of 36 showing less concern about nanotechnology. Pilisuk and Acredolo (1988) show that less educated individuals which are poor, minorities, and among women having higher risk perception toward technology. Additionally, poor people have higher tendency to reject new technology as they perceive high risk in the technology (Boholm 1998). Wealthy people which have the access to education and financially secured have better understanding on science and technology, thus perceive science and technology as beneficial and have the ability to protect them from hazards (Pilisuk and Acredolo 1988). Minorities may be trapped in poverty and have no access to education (Brundtland 1987) and therefore have higher concern about technology (Boholm 1998; Gallup Organization 1979; Vaughan and Nordenstam 1991). Besides that, women show higher risk perception than men since they are more vulnerable, therefore more alert when it comes to risk (Flynn et al. 1994). However, this study did

**Table 2** Significant differences of benefit and risk perceptions among Klang Valley demography

| **Significant difference in benefit perception by demography** | |
|---|---|
| *Race* | |
| Malay | Chinese |
| *Education* | |
| SRP/PMR/LCE | SPM/MCE |
| | Diploma<br>Bachelor degree<br>Master degree |
| **Significant difference in risk perception by demography** | |
| *Age* | |
| 18–20 | 21–40 |
| | 41–51 |
| *Race* | |
| Malay | Indian |
| *Education* | |
| Ph.D | STPM/Matriculation |
| | Diploma |
| | Bachelor degree |
| *Household income* | |
| MYR 1001–RM3000 | MYR 5001–RM7000 |
| MYR 9001 above | MYR 3001–RM5000 |
| | MYR 5001–RM7000 |

not find any significant on different gender in influencing public perception as in previous studies. This inconsistent finding on the demography may in turn suggest that the individual difference in nanotechnology perceptions could not be explained by demographic factors alone. Po et al. (2003) suggest other factors to be considered to further understand public perception. Thus, psychological, sociological, and moderating factors are further discussed in the next section.

## 5.2 Psychological Aspects

Descriptive analyses on the psychological aspects are shown in Table 3 whereby 47.2% of the Klang Valley public has no idea what nanotechnology is. However, they have positive attitude toward nanotechnology. Government, industry, and researchers are the three main stakeholders responsible in developing nanotechnology whereby the public trust shows that researchers are the most trusted stakeholders followed by industry and government.

To further test these psychological aspects in influencing benefit and risk perceptions of nanotechnology, simple regression test is employed and shown in Table 4. The findings are discussed in Sects. 5.2.1, 5.2.2, and 5.2.3.

### 5.2.1 Knowledge

As discussed in the introduction, knowledge influences benefit and risk perceptions. However, the result from this study shows that knowledge has no significant effect on benefit and risk perceptions among the Klang Valley public. The respondents have difficulty in expressing their answer as almost half of them stating that they do not have any knowledge on nanotechnology. However, upon brief explanation given to the respondents, they can answer the survey and express their views and concerns about the risks associated with nanotechnology (Grinbaum 2006). Their perceptions of nanotechnology are carried out heuristically through their knowledge of science and technology, as well as the benefits and risks they derive from different media, including media coverage and product labels (Capon et al. 2015a; Siegrist and Keller 2011).

### 5.2.2 Attitude

Positive public attitude leads to benefit perception of nanotechnology, thereby reduces risk perception of nanotechnology. Attitude functions as a heuristic signal when one remembers the past behavior that will affect their present decision (Bem and McConnell 1970; Pratkanis 1988; Ross et al. 1981). Attitude influences benefit and risk perceptions depending on the public benefit or risk perceived in the

**Table 3** Descriptive analyses on psychological aspects

| Psychological aspects | % | Mean | SD |
|---|---|---|---|
| Knowledge about nanotechnology | 47.2% (know nothing about nanotechnology) | – | – |
| Attitude toward nanotechnology | – | 4.85 | 1.01 |
| Trust in government on nanotechnology development | – | 4.47 | 1.33 |
| Trust in industry on nanotechnology development | – | 4.63 | 1.07 |
| Trust in researchers on nanotechnology development | – | 5.04 | 1.21 |

*Note* Likert scale measurement, 1 = strongly disagree, 7 = strongly agree

**Table 4** Standardized regression coefficient psychological aspects in influencing benefit and risk perceptions of nanotechnology

| Independent variables | | Dependent variables | | | |
|---|---|---|---|---|---|
| | | $R^2$ | Benefit perception (β) | $R^2$ | Risk perception (β) |
| Psychological aspects | Knowledge (independent t-test) | – | -1.140 | – | 1.100 |
| | Attitude | 0.403 | 0.635** | 0.012 | −0.107* |
| | Trust in government | 0.133 | 0.364** | 0.029 | 0.170* |
| | Trust in industry | 0.168 | 0.410** | 0.004 | 0.064 |
| | Trust in researchers | 0.157 | 0.397** | 0.001 | −0.023 |

Significant level at $^{**}p < 0.001$; $^{*}p < 0.05$

nanotechnology applications (Besley 2010; Chen et al. 2013). It is found that public benefit perception on nanotechnology applications outweighs risk perception. The potential of nanotechnology applications in the medical field to improve disease diagnosis and treatment has elevated public benefit perception (Gardner et al. 2010). Meanwhile, concerns on eating nano-related foods which may associate with uncertain health risks raise the public risk perception. The public will have better judgment on nanotechnology when they are familiar with the products (van Giesen et al. 2018). The public familiarity of nanotechnology applications affects their attitude toward the technology and consequently influences their benefit and risk perceptions of nanotechnology (Frewer et al. 2014; Gupta et al. 2015).

### 5.2.3 Trust

Trust is a key factor for public acceptance of particular technology. The public that has the confidence in stakeholders helps the public to accept new technology. The trust that the public has for the government, industry, and researchers increases the public benefit perception of nanotechnology as these stakeholders are the driving forces for the development of nanotechnology. Among the three stakeholders, researchers gain a higher trust from the public in providing information on nanotechnology-related benefits and risks, developing nanotechnology in accordance with the public needs, ensuring the public safety from adverse effects of nanotechnology and having enough technical knowledge in managing nanotechnology development. Lin et al. (2013) agrees that the public has more trust in researchers compared to the government and industry. In addition, this study reveals that the lack of public trust in government compared to the other two stakeholders increases the risk perception among the public, although government is the main regulator that has the authority to manage nanotechnology. The explanation for this circumstance is that the public may have lost the political trust in government due to inefficiency to achieve former policy (Stoker et al. 2017). However, the public does not know that different government agencies are working together to manage nanotechnology risks, where the public relies solely on their belief that the ruling government does not prioritize public needs (Macoubrie 2005). Opposed to the industry and researchers, they are not directly involved in the national policy making, thus gaining greater public trust. The public also has higher confidence for researchers to safeguard public safety from nanotechnology risks attributable to their profound knowledge and skills (Kishimoto 2010).

## 5.3 Sociological Aspects

Descriptive analyses on the sociological aspects are shown in Table 5 whereby the public culture, religious belief, and social aspect are well accepting the nanotechnology.

To further test these sociological aspects in influencing benefit and risk perceptions of nanotechnology, simple regression test is employed and shown in Table 6. The findings are discussed in Sects. 5.3.1, 5.3.2, and5.3.3.

### 5.3.1 Culture

The Malaysian public has been familiar with science and technology on the basis of sociological aspects since the introduction of National Science, Technology and Innovation Policy (2013–2020) highlighting science and technology as the economic driver of the public well-being (Prime Minister's Office 1986). The introduction of this policy has made the development of science and technology in general and nanotechnology in particular to be culturally acceptable to the Klang Valley public; this introduction increases the benefit

**Table 5** Descriptive analyses on sociological aspects

| Sociological aspects | – | Mean | SD |
|---|---|---|---|
| Culturally accept nanotechnology | – | 4.86 | 1.16 |
| Religious beliefs on accepting nanotechnology | – | 4.90 | 1.19 |
| Socially accept research funding of nanotechnology | – | 4.88 | 1.13 |

*Note* Likert scale measurement, 1 = strongly disagree, 7 = strongly agree

**Table 6** Standardized regression coefficient sociological aspects in influencing benefit and risk perceptions of nanotechnology

| Independent variables | | Dependent variables | | | |
|---|---|---|---|---|---|
| | | $R^2$ | Benefit perception (β) | $R^2$ | Risk perception (β) |
| Sociological aspects | Culture | 0.285 | 0.534** | 0.001 | −0.033 |
| | Religious beliefs | 0.364 | 0.604** | 0.002 | −0.045 |
| | Social aspect | 0.354 | 0.595** | 0.002 | −0.041 |

Significant level at $**p < 0.001$; $*p < 0.05$

perception of nanotechnology but not risk perception. Identical to China, the introduction of Open-Door Policy in 1978 which utilized science and technology for improving the standard of living of the public and also enhancing China global competitiveness has made the public to have high acceptance of nanotechnology (Zhang et al. 2015). The introduction of such policy in a country has made the development of science and technology in general and nanotechnology in particular to be culturally acceptable to the public.

### 5.3.2 Religious Beliefs

Religious beliefs influence benefit perception but not risk perception. Religiosity plays a role in influencing ethical choices (Conroy and Emerson 2004; Magill 1992). The development of science and technology emphasizing on religious ethics (Chapman 1999) creates the foundation for accepting nanotechnology for the religious public. Consequently, no technological development conflicting with the ethics practiced in Malaysia, which in turn results in a higher benefit perception of nanotechnology for people with religious beliefs. This result is, however, in contrast with the finding by Brossard et al. (2009) whereby they point out that religious public perceives nanotechnology to be more risky than beneficial.

### 5.3.3 Social Aspect

According to the social implication of nanotechnology on the public, supporting research funding is crucial to ensure fair distribution of nanotechnology benefit while managing the uncertain risk. Thus, social aspect influences benefit perception but not risk perception. Social aspect impacts on society in terms of the development of nanotechnology, enabling the public to engage in nanotechnology development in order to develop nanotechnology that is in line with public interests (Forloni 2012). Public concerns arise from nanotechnology's unbalanced benefits and risks that are not shared equally with all social groups (Conti et al. 2011) whereby social groups with a high standard of living can enjoy the benefits of nanotechnology, while those with a low standard of living will be left behind. Continuous nanotechnology research will therefore enable all social groups in the public to enjoy the benefits and to be protected against the risks of nanotechnology (Kelechukwu 2016).

## 5.4 Other Factors (Moderators)

Other factors are shown to have significant moderating effects on psychological and sociological aspects in influencing public benefit and risk perceptions of nanotechnology. Table 7 shows media coverage, technology and economy development, benefit of nanoapplications, risk of nanoapplications, and benefit and risk information moderate psychological aspects in influencing public perception of nanotechnology, whereas Table 8 shows media coverage, technology and economy development, benefit of nanoapplications, risk of nanoapplications, and benefit and risk information moderate sociological aspects in influencing public perception of nanotechnology. The findings are discussed in Sects. 5.4.1, 5.4.2, 5.4.3, 5.4.4, and 5.4.5.

### 5.4.1 Media

As shown in Table 7, media coverage has a moderating effect on attitude and trust in researchers in influencing benefit perception, while media coverage has a moderating effect on knowledge in influencing risk perception of nanotechnology. Malaysian public shows a positive attitude toward nanotechnology and has high trust in researchers. Media coverage containing useful nanotechnology information on the safety of nanotechnology can further increase the benefit perception of nanotechnology as media is an important medium for keeping the public up to date with the latest information required through researchers' engagement. The information given by the researchers will therefore increase public confidence and further boost public benefit perception. However, when the public is exposed to high media coverage, an increase in risk perception can be observed. This is due to the limited knowledge of nanotechnology among the public. Since 47.2% of the respondents admitting to have zero knowledge about nanotechnology, the risk information become more influential compared to benefit information (Cobb 2005). Consequently, when people with limited knowledge are exposed to too many risk information on the media, they will perceive nanotechnology as risky rather than being beneficial. Media coverage is important for providing information on the benefits and risks of nanotechnology as it has the impact to shape public perceptions toward nanotechnology (Ho et al. 2011). Media coverage gives the public an easy access to information whereby the predominant use of Internet today makes it easy for the public to obtain information. The choice of information, however, depends on the interests and curiosity of individuals. The public interpretation of the received information is also different from one to another (Lemanczyk 2014); the presented information must be factual and non-fictional. Scientific information, however, will only attract certain groups of public who are interested on current scientific developments and leave others with different interest uninformed with the current issue of nanotechnology. Therefore, media covering various aspects including benefits, risks, economics, social, and ethics will provide an extensive coverage to educate and improve the public knowledge (Tyshenko 2014).

### 5.4.2 Technology and Economy Development

As shown in Table 7, technology and economy development moderates knowledge and attitude in influencing benefit perception, while risk perception is moderated by technology and economic development on attitude, trust in government and trust in researchers. As shown in Table 8, technology and economy development moderates culture and social aspect in influencing risk perception but shows no moderating effect in influencing benefit perception. As technology and economy continue to develop, the public will gain better understanding and familiarity with the new technology that will increase public knowledge and positive attitude toward nanotechnology. An economy driven by the advancement of science and technology will allow the competitiveness among countries to be part of the Fourth Industrial Revolution (Tangau 2017). Therefore, in conjunction with technology and economic development, public culture and social aspect agree on science and technology to fuel the growth of economy and enhance the public well-being, resulting in a reduction of public risk perception of nanotechnology. Countries that have never had terrible experience of scientific and technological development, such as technological disasters, would not lead to culture and social phobia in adopting new technologies, such as nanotechnology (Macnaghten et al. 2016; Roco and Bainbridge, 2001). Public that is protected from any controversial development of technology will accept new technologies by trusting the government and researchers which result in the decrease in risk perception among them. Trust is noteworthy in a stable technological and economic development, as it significantly influences public perception to illustrate whether nanotechnology will be accepted or rejected.

### 5.4.3 Benefit of Nanoapplications

As shown in Table 7, benefit perceived in nanoapplications shows a moderating effect on trust in government and industry in influencing benefit perception, while moderating attitude, trust in government, trust in industry, and trust in researchers in influencing risk perception. Table 8 shows the sociological aspects in which culture, religious beliefs, and social aspect are moderated by perceived benefit of nanoapplications in influencing risk perception of nanotechnology. It is found that the perceived benefit of nanoapplications helps to contribute to public confidence in government and industry which increases benefit perception of nanotechnology, as both government and industry play a critical role in the management and distribution of useful nanotechnology products to the public. This finding is further supported by Maynard (2006) that public has higher trust in government and industry when nanotechnology applications are beneficial. In addition, perceived benefit in nanoapplications is found to be affecting attitude, trust in government, trust in industry, and trust in researchers, resulting in the decrease of public risk perception. The public finds that nanotechnology is beneficial; thus, their attitude toward nanotechnology is also positive. Consequently, the public has a low risk perception of nanotechnology. Public trust in government, trust in industry, and

**Table 7** Regression coefficients from PROCESS macro by Hayes for the moderating effects of psychological aspects in influencing the public perception of nanotechnology

| Psychological aspects | | Media coverage | Technology and economy development | Benefit of nanoapplications | Risk of nanoapplications | Benefit and risk information |
|---|---|---|---|---|---|---|
| Benefit perception | Knowledge | −0.11 | 0.28* | −0.01 | 0.25 | 0.01 |
| | Attitude | 0.05* | 0.10** | 0.03 | −0.03 | 0.01 |
| | Trust in government | 0.03 | −0.04 | 0.09** | −0.04 | −0.02 |
| | Trust in industry | −0.01 | 0.03 | 0.05* | −0.03 | −0.02 |
| | Trust in researchers | 0.05* | 0.02 | 0.02 | −0.08* | −0.05* |
| Risk perception | Knowledge | 0.53** | 0.17 | 0.13 | −0.01 | 0.30* |
| | Attitude | −0.07 | −0.08* | −0.25** | 0.03 | −0.02 |
| | Trust in government | −0.04 | −0.01** | 0.17** | 0.07** | −0.07* |
| | Trust in industry | −0.02 | −0.06 | −0.18** | 0.07** | 0.11* |
| | Trust in researchers | 0.03 | −0.09* | −0.18** | 0.05* | 0.14** |

Significant level at $^{**}p < 0.001$; $^{*}p < 0.05$

**Table 8** Regression coefficients from PROCESS macro by Hayes for the moderating effects of sociological aspects in influencing the public perception of nanotechnology

| Sociological aspects | | Media coverage | Technology and economy development | Benefit of nanoapplications | Risk of nanoapplications | Benefit and risk information |
|---|---|---|---|---|---|---|
| Benefit perception | Culture | 0.04 | 0.02 | 0.04 | −0.01 | 0.01 |
| | Religious beliefs | 0.03 | 0.04 | −0.02 | −0.08** | 0.01 |
| | Social aspect | 0.03 | 0.03 | 0.01 | −0.05* | 0.00 |
| Risk perception | Culture | −0.02 | −0.12* | −0.24** | 0.03 | −0.04 |
| | Religious beliefs | 0.02 | −0.07 | −0.23** | 0.03 | −0.02 |
| | Social aspect | −0.01 | −0.10* | −0.26** | 0.03 | −0.03 |

Significant level at $^{**}p < 0.001$; $^{*}p < 0.05$

trust in researchers are vital which may boost public confidence to perceive benefit in nanoapplications, where in this manner, leading their risk perception to be diminished (Capon et al. 2015b). Furthermore, public culture, religious beliefs, and social aspect which are positive toward nanotechnology, resulting in public tendency to perceive benefit in nanoapplications, thereby reduce their risk perception of nanotechnology (Mamadouh 1999).

### 5.4.4 Risk of Nanoapplications

As shown in Table 7, risk of nanoapplications is found to have moderating effect on psychological aspects whereby it moderates trust in researchers in influencing benefit perception of nanotechnology. At the same time, risk of nanoapplications moderates trust in government, trust in industry, and trust in researchers in influencing risk perception of nanotechnology. For sociological aspects, Table 8 indicates religious beliefs and social aspect are moderated by risk perceived in nanoapplications in influencing benefit perception of nanotechnology. It is an interesting finding where high public trust in researchers leads to the increase of benefit perception, although there are risks associated with nanotechnology applications. The public is confident that researchers will protect them from nanotechnology risks and thus increase their benefit perception toward nanotechnology (Kishimoto 2010). Nevertheless, if too many risks are associated with nanotechnology applications, this negativity will affect the public confidence in researchers, thereby reducing their benefit perception. Very high risk perceived from nanoapplications, on the other hand, increases risk perception and in a long run may deprive public trust in government, industry, and researchers (Oh 2009). It is therefore imperative for government, industry, and researchers to manage nanotechnology risks in order to avoid public trust deprivation that may inhibit the development of nanotechnology. It is also shown that public religious belief and social aspect are positive about nanotechnology, thus perceiving nanotechnology as beneficial although there are risks associated with nanoapplications. People who adhere to religion believe that it is important to make ethical choices including the safety application of nanotechnology in consumer products, given that religious beliefs act as guidance for them to choose ethically (Conroy and Emerson 2004; Magill 1992).

### 5.4.5 Benefit and Risk Information

As shown in Table 7, benefit and risk information shows a moderating effect on trust in researchers in influencing benefit perception. Benefit and risk information also moderates knowledge, trust in government, trust in industry, and trust researchers in influencing risk perception. However, there is no significant moderating effect by benefit and risk information on sociological aspects as shown in Table 8. The benefit and risk information in the form of nanoproduct label serves as a communication tool for public decision making (Chuah et al. 2018). Public relies on their trust in researchers which result in increasing benefit perception when there is a lack of nanotechnology information available. Public trust in researchers is crucial in influencing benefit perception, whereby the public is willing to accept vulnerability when they have a high positive expectation for researchers (Roosen et al. 2015). Knowledge, on the other hand, shows to increase risk perception although benefit and risk information is readily available to the public. This situation is caused by limited knowledge about nanotechnology among the public. As risk information is more influential, public with a different background is going to interpret the same information differently (Douglas 1978). The balance of benefit and risk information disclosed to the public is critical in order to avoid misinterpretation of information, as the public may not have the expertise and may be attracted to risk information more than benefit information (Siegrist and Keller 2011). In conjunction with information availability, lack of information provided to the public can increase the risk perception of nanotechnology and may deteriorate public trust in government, industry, and researchers. Public conveys the need for information from the expert to reduce risk perception in the midst of nanotechnology uncertainties. Therefore, mandatory labeling is required to gain public trust and reduce public concern regarding nanotechnology risks (Forloni 2012). In addition, adequate information will enable the public to make decisions objectively without relying excessively on their trust in government, industry, and researchers which may be biased and could lead to the wrong decision (Gilovich et al. 2002).

## 6 Conclusions

Nanotechnology is one of the most advanced technologies used in facing Fourth Industrial Revolution. The development of nanotechnology from researches in laboratories has moved to the commercialization of the nanotechnology products in the market. The public are consumers of nanotechnology products that will receive impacts from products containing nanomaterials. Good quality, durable, anti-bacterial, and anti-fungal products will increase consumers' interest in using nano-containing products. The advancement of nanotechnology use in medicine has improved the diagnosis and treatment of diseases that would benefit the public in improving public health. However, nanotechnology also has uncertain risks that can bring health problems to the public and pollute the environment. Nanotechnology is still at its early stage of development in Malaysia, so the public perceptions of nanotechnology will be instrumental for policy-makers to develop nanotechnology so that its benefits can be enjoyed equally by all levels of society and the risks can be well managed.

In general, public perceptions of nanotechnology in Malaysia are positive as Malaysia receives both scientific and technological developments as drivers for the country's economic growth. Public perceptions based on psychological aspects show that knowledge is not a factor affecting the benefit and risk perceptions. However, the public perceives nanotechnology to be more beneficial than risky. Public attitudes are also positive for nanotechnology in which this positive attitude gives people benefit perception and reduces risk perception of nanotechnology. Trust in government, industry, and researchers increase the public benefit perception on nanotechnology as these stakeholders are the driving force of the development of nanotechnology in Malaysia. The government as the regulator of nanotechnology development affects risk perception when public trust in government declines. Therefore, the government needs to play a transparent role in getting public trust, thereby enhancing the public benefit perception on nanotechnology.

Based on sociological aspects, culture, religious belief, and social aspect in Malaysia influence the public benefit perception but not risk perception on nanotechnology. The National Science and Technology Policy has emphasized science and technology as an economic driver for the well-being of Malaysian. Hence, it has become a culture in Malaysia receiving both scientific and technological developments in general and nanotechnology in particular. Religions in Malaysia accept nanotechnology whereby technology development in Malaysia must be parallel with Islam, the official religion of the nation, which clearly rejects any form of technology that violates ethics. Hence, no ethical conflicting technology can be practiced in Malaysia. Correspondingly, continuous research of nanotechnology will result in the social implication by ensuring equal distribution of nanotechnology benefit and at the same time its risk will be effectively managed. Thus, social aspect does effect on the benefit perception and does not affect the risk perception of nanotechnology.

There are other factors (moderators), which are media coverage, technology and economy development, benefit of nanotechnology applications, risk of nanotechnology applications and benefit and risk information, moderating psychological and sociological aspects in influencing public perception of nanotechnology. Based on psychological aspects, media coverage moderates attitudes and trust in researchers in influencing benefit perception of nanotechnology, whereas risk perception of nanotechnology is influenced by the moderating effect of media coverage on knowledge. However, the media coverage does not affect the sociological aspects. Media being an important medium for communicating information to the public requires the involvement of researchers so that nanotechnology can be communicated to the public effectively. Hence, avoiding risk information dominating the media coverage without supporting fact will increase public risk perception, especially to those with limited knowledge about nanotechnology. A comprehensive media coverage covering various areas about nanotechnology will help stakeholders to create awareness to all levels of society which comprises various interests, propositions, and backgrounds.

Based on psychological aspects, rapid technology and economic developments have the moderating effect on public knowledge and attitudes which increase the benefit perception of nanotechnology. On the other hand, attitudes' trust in government and researchers reduce risk perception of nanotechnology as technology and economy continuously developed. Based on sociological aspects, nanotechnology which is accepted by public culture and social status reduces risk perception as a result of advanced technology and economic developments. Numerous studies on nanotechnology have illustrated its growing development and gradually impacting the economy. Therefore, this study suggests that public must also be informed of recent studies so that the public can obtain factual information from researchers. It also reduces the gap between researchers and the public by encouraging the public to make scientific decisions so that their responses to express the needs and wants from nanotechnology development are factual. The effectiveness of public communication will facilitate policy-makers to develop nanotechnology that is in line with the public needs well guided rather than just mere emotion.

Based on psychological aspects, the benefit of nanotechnology applications is affecting trust in government and industry that influence public benefit perception. Both stakeholders play a role in managing and delivering useful products to the public. By perceiving benefit of nanotechnology applications, public positive attitude, their trust in government, industry, and researchers will reduce the risk perception of nanotechnology. As for sociological aspects, public culture, religious beliefs, and social aspect which well accept nanotechnology have prompted them to find nanotechnology as beneficial, causing risk perception of nanotechnology to be diminished.

For the risk of nanotechnology applications based on psychological aspects, public has high trust in researchers which enables them to perceive benefit despite the risk of nanotechnology applications. However, when they perceive risk of nanotechnology applications is too high, the public trust in government, industry, and researchers will increase their risk perception. Based on sociological aspects, public religious beliefs and social aspect accept nanotechnology well resulting in benefit perception despite the risk associated with nanotechnology applications. However, high risk will still reduce their benefit perception of nanotechnology. If the risk of nanotechnology applications is poorly managed, it will cause the deterioration of public trust of the stakeholders to develop nanotechnology in Malaysia. Consequently, it is proposed that the benefit and risk of nanotechnology applications shall be informed to the public through continuous research on products containing nanomaterials. As a result, public trust in stakeholders will increase whereby public is more confident on the benefit of nanotechnology that they receive and protect them from the unwanted risk.

The benefit and risk information on the product acts as a guide for the public to make choices. Based on psychological aspects, benefit and risk information has the effect on public trust in researchers which further increase public benefit perception of nanotechnology, whereas for risk perception, adequate information availability has the effect on public trust in government, industry, and researchers which has potential in reducing risk perception. However, different circumstances are found in knowledge. Comprehensive benefit and risk information, however, increases the risk perception among the public; whereby with the public limited knowledge about nanotechnology, risk information is more likely to influence public perception than benefit information. However, there is no significant moderating effect by the benefit and risk information on sociological aspects in influencing public perception of nanotechnology. Benefit and risk information of nanotechnology on the product is an important tool for the public to make decision. Labeling a product containing nanomaterials with both risk and benefit information supported by reliable research evidences is urgently needed as nanotechnology products have already entered the consumer market. Finally, mandatory law enforcement on labeling is required to promptly increase public trust in stakeholders and subsequently develop nanotechnology to drive sustainable development in ensuring the safety and well-being of the public.

## References


Allsopp, M., Walters, A. & Santillo, D., (2007). *Nanotechnologies and nanomaterials in electrical and electronic goods: a review of uses and health concerns. Greenpeace Research Laboratories*, London, viewed 7 April 2017, http://www.nanometer.ru/2007/12/26/nanomaterial_5521/PROP_FILE_files_1/nanotech_in_electronics_2007.pdf.

Bainbridge, W. S. (2002). Public attitudes toward nanotechnology. *Journal of Nanoparticle Research, 4*(6), 561–570. https://doi.org/10.1023/A:1022805516652.

Bem, D. J., & McConnell, H. K. (1970). Testing the self-perception explanation of dissonance phenomena: On the salience of premanipulation attitudes. *Journal of Personality and Social Psychology, 14*(1), 23–31. https://doi.org/10.1037/h0020916.

Bennet-Woods, D., (2008). *Nanotechnology: Ethics and society*. M. T. Burke, G. L. Hornyak, D. Bennet-Woods, J. A. Shatkin, & P. M. Bouche (Eds.), New York: Taylor & Francis Group.

Besley, J. (2010). Current research on public perceptions of nanotechnology. *Emerging Health Threats Journal, 3*(1), 7098. https://doi.org/10.3134/ehtj.10.164.

Binder, A. R., Cacciatore, M. A., Scheufele, D. A., Shaw, B. R., & Corley, E. A. (2012). Measuring risk/benefit perceptions of emerging technologies and their potential impact on communication of public opinion toward science. *Public Understanding of Science, 21*(7), 830–847. https://doi.org/10.1177/0963662510390159.

Boholm, A. (1998). Comparative studies of risk perception: A review of twenty years of research. *Journal of Risk Research, 1*(2), 135–163.

Bostrom, A., & Löfstedt, R. E. (2010). Nanotechnology risk communication past and prologue. *Risk Analysis: An International Journal, 30*(11), 1645–1662. https://doi.org/10.1111/j.1539-6924.2010.01521.x.

Brossard, D., Scheufele, D. A., Kim, E., & Lewenstein, V. (2009). Religiosity as a perceptual filter: Examining processes of opinion formation about nanotechnology. *Public Understanding of Science, 18*(5), 546–558. https://doi.org/10.1177/0963662507087304.

Brown, J., & Kuzma, J. (2013). Hungry for information: public attitudes toward food nanotechnology and labeling. *Review of Policy Research, 30*(5), 512–548.

Brundtland, G. H. (1987). Our common future: Report of the world commission on environment and development. *United Nations Commission, 4*(1), 300. https://doi.org/10.1080/07488008808408783.

Burri, R. V., & Bellucci, S. (2008). Public perception of nanotechnology. *Journal of Nanoparticle Research, 10*(3), 387–391. https://doi.org/10.1007/s11051-007-9286-7.

Cacciatore, M. A., Scheufele, D. A., & Corley, E. A. (2011). From enabling technology to applications: The evolution of risk perceptions about nanotechnology. *Public Understanding of Science, 20* (3), 385–404. https://doi.org/10.1177/0963662509347815.

Capon, A., Gillespie, J., Rolfe, M., & Smith, W. (2015a). Comparative analysis of the labelling of nanotechnologies across four stakeholder groups. *Journal of Nanoparticle Research, 17*(237), 1–13. https://doi.org/10.1007/s11051-015-3129-8.

Capon, A., Gillespie, J., Rolfe, M., & Smith, W. (2015b). Perceptions of risk from nanotechnologies and trust in stakeholders: A cross sectional study of public, academic, government and business attitudes. *BMC Public Health, 15*(1), 424. https://doi.org/10.1186/s12889-015-1795-1.

Chapman, A. R. (1999). *Unprecedented choices: Religious ethics at the frontier of genetic science*. Minneapolis: Fortress Press.

Chen, M. F., Lin, Y. P., & Cheng, T. J. (2013). Public attitudes toward nanotechnology applications in Taiwan. *Technovation, 33*(2–3), 88–96. https://doi.org/10.1016/j.technovation.2012.11.008.

Chuah, A. S. F., Leong, A. D., Cummings, C. L., & Ho, S. S. (2018). Label it or ban it? Public perceptions of nano-food labels and propositions for banning nano-food applications. *Journal of Nanoparticle Research, 2,* 1–17.

Cobb, M. D. (2005). Framing effects on public opinion about nanotechnology. *Science Communication, 27*(2), 221–239. https://doi.org/10.1177/1075547005281473.

Cobb, M. D., & Macoubrie, J. (2004). Public perceptions about nanotechnology: Risks, benefits and trust. *Journal of Nanoparticle Research, 6*(4), 395–405. https://doi.org/10.1007/s11051-004-3394-4.

Conroy, S. J., & Emerson, T. L. N. (2004). Ethics and religion: As a predictor of religiosity business ethical awareness students. *Journal of Business Ethics, 50*(4), 383–396. https://doi.org/10.1023/B:BUSI.0000025040.41263.09.

Conti, J., Satterfield, T., & Harthorn, B. H. (2011). Vulnerability and social justice as factors in emergent U. *S. Nanotechnology Risk Perceptions. Risk Analysis, 31*(11), 1734–1748. https://doi.org/10.1111/j.1539-6924.2011.01608.x.

Corley, E. A., & Scheufele, D. A. (2010). Outreach going wrong? When we talk nano to the public, we are leaving behind key audiences. *The Scientist, 24*(1), 22. https://doi.org/10.1086/266828.

Cormick, C. (2009). Why do we need to know what the public thinks about nanotechnology? *NanoEthics, 3*(2), 167–173. https://doi.org/10.1007/s11569-009-0065-z.

Currall, S. C. (2009). New insights into public perceptions. *Nature Nanotechnology, 4,* 79–80. https://doi.org/10.1038/nnano.2008.423.

Damasio, A. R., Tranel, D., & Damasio, H. (1990). Individuals with sociopathic behavior caused by frontal damage fail to respond autonomically to social stimuli. *Behavioural Brain Research, 41,* 81–94. https://doi.org/10.3906/mat-1703-92.

De Luca, A., & Ferrer, B. B. (2017). Nanomaterials for water remediation: Synthesis, application and environmental fate. *Nanotechnologies for environmental remediation* (pp. 25–60). New York: Springer, Cham.

Dijkstra, A. M., & Critchley, C. R. (2014). Nanotechnology in Dutch science cafes: Public risk perceptions contextualised. *Public Understanding of Science, 25*(1), 71–87. https://doi.org/10.1177/0963662514528080.

Douglas, M. (1978). Cultural bias: Royal anthropological institute of great Britain and Ireland. *Occasional Paper, 35,* 1978. https://doi.org/10.1017/CBO9781107415324.004.

Finucane, M. L., Alhakami, A. L. I., Slovic, P., & Johnson, S. M. (2000). The affect heuristic in judgments of risks and benefits. *Journal of Behavioral Decision Making, 3,* 1–17.

Fischhoff, B., Slovic, P., & Lichtenstein, S. (1983). "The Public" Vs. "The Experts": Perceived Vs. Actual Disagreements About Risks of Nuclear Power. In *The analysis of actual versus perceived risks* (pp. 235–249). Boston, MA: Springer.

Flynn, J., Slovic, P., & Mertz, C. K. (1994). Gender, race, and perception of environmental health risks. *Risk Analysis, 14*(6), 1101–1108. https://doi.org/10.1021/acsmacrolett.6b00822.

Forloni, G. (2012). Responsible nanotechnology development. *Journal of Nanoparticle Research, 14*(8), 1–17. https://doi.org/10.1007/s11051-012-1007-1.

Frewer, L. J., Gupta, N., George, S., Fischer, A. R. H., Giles, E. L., & Coles, D. (2014). Consumer attitudes towards nanotechnologies applied to food production. *Trends in Food Science & Technology, 40*(2), 211–225. https://doi.org/10.1016/j.tifs.2014.06.005.

Fromer, N. A., & Diallo, M. S. (2013). Nanotechnology and clean energy: Sustainable utilization and supply of critical materials. *Journal of Nanoparticle Research, 15,* 289–304. https://doi.org/10.1007/978-3-319-05041-6_23.

Gallup Organization. (1979). *Nuclear power plant*. New York: Gallup Report.

Gardner, G., Jones, G., Taylor, A., Forrester, J., & Robertson, L. (2010). Students' risk perceptions of nanotechnology applications: Implications for science education. *International Journal of Science Education, 32*(14), 1951–1969.

Gaskell, G., Eyck, T. Ten, Jackson, J., & Veltri, G. (2005). Imagining nanotechnology: Cultural support for technological innovation in Europe and the United States. *Public Understanding of Science, 14,* 81–90. https://doi.org/10.1177/0963662505048949.

George, S., Kaptan, G., Lee, J., & Frewer, L. (2014). Awareness on adverse effects of nanotechnology increases negative perception among public: Survey study from Singapore. *Journal of Nanoparticle Research, 16*(2751), 1–11. https://doi.org/10.1007/s11051-014-2751-1.

Giles, E. L., Kuznesof, S., Clark, B., Hubbard, C., & Frewer, L. J. (2015). Consumer acceptance of and willingness to pay for food nanotechnology: A systematic review. *Journal of Nanoparticle Research, 17* (12), 1–26. https://doi.org/10.1007/s11051-015-3270-4.

Gilovich, T., Griffin, D., & Kahneman, D. (2002). *Heuristics and biases: The psychology of intuitive judgment.* In T. Gilovich, D. Griffin, & D. Kahneman, (Eds.), *Academy of management review*, Cambridge: Cambridge University Press.

Gleiche, M., Hoffschulz, H., & Lenhert, S. (2006). *Nanotechnology in consumer products*, viewed 16 July 2017 https://www.nanowerk.com/nanotechnology/reports/reportpdf/report64.pdf.

Gorss, J. B. (2008). *Framing nano: Media coverage and public opinion about nanotechnology*, M.A. Dissertation: Cornell University, Ithaca, New York, viewed 16 May 2017, http://citeseerx.ist.psu.edu/viewdoc/download?doi=10.1.1.854.6844&rep=rep1&type=pdf.

Grinbaum, A. (2006). Cognitive barriers in perception of nanotechnology. *Journal of Law, Medicine and Ethics, 34*(4), 689–694. https://doi.org/10.1111/j.1748-720X.2006.00088.x.

Gupta, N., Fischer, A. R. H., & Frewer, L. J. (2015). Ethics, risk and benefits associated with different applications of nanotechnology: A comparison of expert and consumer perceptions of drivers of societal acceptance. *NanoEthics, 9*(2), 93–108. https://doi.org/10.1007/s11569-015-0222-5.

Gupta, N., Fischer, A. R. H., Van Der Lans, I. A., & Frewer, L. J. (2012). Factors influencing societal response of nanotechnology:

An expert stakeholder analysis. *Journal of Nanoparticle Research, 14*(5), 1–15. https://doi.org/10.1007/s11051-012-0857-x.

Handford, C. E., Dean, M., Henchion, M., Spence, M., Elliott, C. T., & Campbell, K. (2014). Implications of nanotechnology for the agri-food industry: Opportunities, benefits and risks. *Trends in Food Science & Technology, 40*(2), 226–241. https://doi.org/10.1016/j.tifs.2014.09.007.

Harifi, T., & Montazer, M. (2017). Application of nanotechnology in sports clothing and flooring for enhanced sport activities, performance, efficiency and comfort: A review. *Journal of Industrial Textiles, 46*(5), 1147–1169. https://doi.org/10.1177/1528083715601512.

Hashim, U., Nadia, E., & Salleh, S. (2009). Nanotechnology development status in Malaysia: industrialization strategy and practices. *Int. J. Nanoelectronics and Materials, 2*(1), 119–134.

Ho, S. S., Scheufele, D. A., & Corley, E. A. (2011). Value predispositions, mass media, and attitudes toward nanotechnology: The Interplay of public and experts. *Science Communication, 33*(2), 167–200. https://doi.org/10.1177/1075547010380386.

Hristozov, D., & Malsch, I. (2009). Hazards and risks of engineered nanoparticles for the environment and human health. *Sustainability, 1*(4), 1161–1194. https://doi.org/10.3390/su1041161.

Hurni, H., & Wiesmann, U. (2014). Transdisciplinarity in practice: Experience from a concept-based research programme addressing global change and sustainable development. *GAIA—Ecological Perspectives for Science and Society*, *23*(3), 275–277. https://doi.org/10.14512/gaia.23.3.15.

Kahan, D. M., Braman, D., Slovic, P., Gastil, J., & Cohen, G. (2009). Cultural cognition of the risks and benefits of nanotechnology. *Nature Nanotechnology, 4*(February), 2–5. https://doi.org/10.1038/NNANO.2008.341.

Kamarulzaman, N. A., Lee, K. E., & Siow, K. S. (2018). Public perception of nanotechnology for good governance: A conceptual framework for psychological and sociological approaches. *Journal of Food, Agriculture and Environment, 16*(2), 168–174.

Kamarulzaman, N. A., Lee, K. E., Siow, K. S., & Mokhtar, M. (2019). Psychological and sociological persepctives for good governance of sustainable nanotechnology development in Malaysia. *Journal of Nanoparticle Research, 21*(7), 164.

Kass, G. (2001). Open channels: Public dialogue in science and technology (153).

Kelechukwu, E. (2016). Social, legal, ethical, health, safety and environmental aspects of nanotechnology. Viewed 5 January 2017 http://www.academia.edu/26150828/SOCIAL_LEGAL_ETHICAL_HEALTH_SAFETY_AND_ENVIROMENTAL_ASPECTS_OF_NANOTECHNOLOGY.

Kharat, M. G., Murthy, S., & Kamble, S. J. (2017). Environmental applications of nanotechnology: A review. *ADBU Journal of Engineering Technology*, *6*(3).

Kim, Y. (2017). *The role of science in nanotechnology decision-making : Toward evidence-based policy making*, Ph.D. Thesis: Arizona State University, viewed 20 April 2018, https://repository.asu.edu/attachments/191156/content/Kim_asu_0010E_17291.pdf.

Kishimoto, A. (2010). *Public perception of nanotechnologies in Japan from 2005 to 2009*. Japan.

Könninger, S., Ott, I., Zulsdorf, T., & Papilloud, C. (2010). Public reactions to the promotion of nanotechnologies in society. *International Journal of Nanotechnology, 7*(2–3), 265–279.

Lee, C. J., Lee, S., Jhon, M. S., & Shin, J. (2013). Factors influencing nanotechnology commercialization: An empirical analysis of nanotechnology firms in South Korea. *Journal of Nanoparticle Research, 15*(2), 1444.

Lee, C.-J., Scheufele, D. A., & Lewenstein, B. V. (2005). Public attitudes toward emerging technologies: Examining the interactive effects of cognitions and affect on public attitudes toward nanotechnology. *Science Communication, 27*(2), 240–267. https://doi.org/10.1177/1075547005281474.

Leinonen, A., & Kivisaari, S. (2010). Nanotechnology perceptions: Literature review on media coverage, public opinion and NGO perspectives. VTT.

Lemanczyk, S. (2014). Science and national pride: The iranian press coverage of. *Science Communication, 36*(2), 194–218. https://doi.org/10.1177/1075547013516873.

Leung, Y. (2007). *Encyclopedia of behavioral medicine*. New York: Springer Science + Business Media.

Lima, M. L., Barnett, J., & Vala, J. (2005). Risk perception and technological development at a societal level. *Risk Analysis, 25*(5), 1229–1239. https://doi.org/10.1111/j.15396924.2005.00664.x.

Lin, S. F., Lin, H. S., & Wu, Y. Y. (2013). Validation and exploration of instruments for assessing public knowledge of and attitudes toward nanotechnology. *Journal of Science Education and Technology, 22*(4), 548–559. https://doi.org/10.1007/s10956-012-9413-9.

Liu, X., Zhang, P., Li, X., Chen, H., Dang, Y., Larson, C., et al. (2009). Trends for nanotechnology development in China, Russia, and India. *Journal of Nanoparticle Research, 11*(8), 1845–1866. https://doi.org/10.1007/s11051-009-9698-7.

Macnaghten, P., Kearnes, M. B., & Wynne, B. (2016). Nanotechnology, governance and public deliberation: What role for the social sciences? *Science Communication, 27*(2), 268–291. https://doi.org/10.1177/1075547005281531.

Macoubrie, J. (2005). *Informed Public Perceptions of Nanotechnology and Trust in Government. Woodrow Wilson International Center for Scholars*, viewed 10 June 2016, https://www.wilsoncenter.org/sites/default/files/macoubriereport1.pdf.

Macoubrie, J. (2006). Nanotechnology: Public concerns, reasoning and trust in government. *Public Understanding of Science, 15,* 221–241. https://doi.org/10.1177/0963662506056993.

Magill, G. (1992). Theology in business ethics: Appealing to the religious imagination. *Journal of Business Ethics, 11*(2), 129–135. https://doi.org/10.1007/BF00872320.

Mamadou, S. D., Fromer, N. A., & Jhon, M. S., (2012). Nanotechnology for sustainable development. *Journal of Nanoparticle Research, 14.* https://doi.org/10.1007/978-3-319-05041-6.

Mamadouh, V. (1999). Grid-group cultural theory: An introduction. *GeoJournal, 47*(3), 395–409. https://doi.org/10.1023/A:1007024008646.

Masrom, A. K. (2012). National nanotechnology directorate—driving nanomalaysia agenda towards 2020, viewed 15 July 2016, https://docplayer.net/15341022-National-nanotechnology-directorate-driving-nanomalaysia-agenda-towards-2020.html.

Maynard, A. D. (2006). Nanotechnology: Assessing the risks. *Nano Today, 1*(2), 22–33.

Maynard, A. D. (2015). Navigating the fourth industrial revolution. *Nature Nanotechnology, 10*(12), 1005–1006. https://doi.org/10.1038/nnano.2015.286.

Mehic, S. H. (2012). Application of nanotechnology in synthetic detergents production, Republic of Slovenia.

Mensch, F., & Umwelt. (2014). Use of nanomaterials in energy storage, viewed 10 June 2016. https://www.umweltbundesamt.de/sites/default/files/medien/376/publikationen/use_of_nanomaterials_in_energy_storage.pdf.

Metag, J., & Marcinkowski, F. (2014). Technophobia towards emerging technologies? *A comparative analysis of the media coverage of nanotechnology in Austria, Switzerland and Germany, Journalism, 15*(4), 463–481. https://doi.org/10.1177/1464884913491045.

Michelson, E. S. & Rejeski, D. (2006). Falling through the cracks? Public perception, risk, and the oversight of emerging nanotechnologies woodrow wilson international center for david rejeski woodrow wilson international center for. IEEE, pp. 0–16.

Moussaouy, A. El. (2018). *Environmental Nanotechnology and Education for Sustainability: Recent progress and perspective, handbook of environmental materials management* (pp. 1–27). New York: Springer, Cham.

Mu, L., & Sprando, R. L. (2014). Application of nanotechnology in cosmetic. *Research Journal of Pharmacy and Technology, 7*(1), 81–83. https://doi.org/10.1007/s11095-010-0139-1.

Musazzi, U. M., Marini, V., Casiraghi, A., & Minghetti, P. (2017). Is the European regulatory framework sufficient to assure the safety of citizens using health products containing nanomaterials? *Drug Discovery Today, 22*(6), 870–882. https://doi.org/10.1016/j.drudis.2017.01.016.

Nisbet, M. C. & Huge, M. (2007). Where do science debates come from? understanding attention cycles and framing. *The Public, The Media & Agricultural Biotechnology*, pp. 193–230.

Oh, S. H. (2009). Perceptions of nanotechnology in Canada and South Korea, M.A. Dissertation, University of Manitoba Winnipeg, Manitoba, Canada.

Oltedal, S., Moen, B. E., Klempe, H., & Rundmo, T. (2004). Explaining risk perception: An evaluation of cultural theory. *Trondheim: Norwegian University of Science and Technology, 85* (1–33), 86. https://doi.org/10.1080/135753097348447.

Parisi, C., Vigani, M., & Rodríguez-cerezo, E. (2014). Agricultural Nanotechnologies: What are the current possibilities ? *Nano Today*, 10–13. https://doi.org/10.1016/j.nantod.2014.09.009.

Petersen, A., Anderson, A., Wilkinson, C., & Allan, S. (2007). Nanotechnologies, risk and society. *Health, Risk & Society, 9*(2), 117–124. https://doi.org/10.1080/13698570701306765.

Phoenix, C., & Treder, M. (2003). Safe utilization of advanced nanotechnolog, *Center for Responsible Nanotechnology*, pp. 1–10, viewed 9 June 2016, http://citeseerx.ist.psu.edu/viewdoc/download?doi=10.1.1.98.2829&rep=rep1&type=pdf.

Piccinno, F., Gottschalk, F., Seeger, S., & Nowack, B. (2012). Industrial production quantities and uses of ten engineered nanomaterials in Europe and the world. *Journal of Nanoparticle Research, 14*(9), 1109. https://doi.org/10.1007/s11051-012-1109-9.

Pidgeon, N., Harthorn, B. H., Bryant, K., & Rogers-Hayden, T. (2009). Deliberating the risks of nanotechnologies for energy and health applications in the United States and United Kingdom. *Nature Nanotechnology, 4*(2), 95–98. https://doi.org/10.1038/nnano.2008.362.

Pieper, M. H. (1989). The heuristic paradigm: A unifying and comprehensive approach to social work research. *Smith College Studies in Social Work, 60*(1), 8–34. https://doi.org/10.1080/00377318909516663.

Pilisuk, M., & Acredolo, C. (1988). Fear of technological hazards: One concern or many? *Social Behaviour, 3*(1), 17–24.

Po, M., Kaercher, J. D., & Nancarrow, B. E. (2003). Literature review of factors influencing public perceptions of water reuse. *CSIRO Land and Water Technical Report, 54*(03), 1–44.

Pratkanis, A. R. (1988). The attitude heuristic and selective fact identification. *British Journal of Social Psychology, 27*(3), 257–263. https://doi.org/10.1111/j.2044-8309.1988.tb00827.x.

Prime Minister's Office. (1986). National science technology policy, viewed 9 June 2016. http://www.mosti.gov.my/index.php?option=com_content&view=article&id=2032&lang=bm.

Raffa, V., Vittorio, O., Riggio, C., & Cuschieri, A. (2010). Progress in nanotechnology for healthcare. *Minimally Invasive Therapy and Allied Technologies, 19*(3), 127–135. https://doi.org/10.3109/13645706.2010.481095.

Raj, S., Sumod, U., Jose, S., & Sabitha, M. (2012). Nanotechnology in cosmetics: Opportunities and challenges. *Journal of Pharmacy and Bioallied Sciences, 4*(3), 186. https://doi.org/10.4103/0975-7406.99016.

Renn, O., & Roco, M. C. (2006). Nanotechnology and the need for risk governance. *Journal of Nanoparticle Research, 8*(2), 153–191. https://doi.org/10.1007/s11051-006-9092-7.

Renn, O., & Swaton, E. (1984). Psychological and sociological approaches to study risk perception. *Environment International, 10*, 557–575.

Retzbach, A., Marschall, J., Rahnke, M., Otto, L., & Maier, M. (2011). Public understanding of science and the perception of nanotechnology: The roles of interest in science, methodological knowledge, epistemological beliefs, and beliefs about science. *Journal of Nanoparticle Research, 13*(12), 6231–6244. https://doi.org/10.1007/s11051-011-0582-x.

Rist, S., Chidambaranathan, M., Escobar, C., Wiesmann, U., & Zimmermann, A. (2007). Moving from sustainable management to sustainable governance of natural resources: The role of social learning processes in rural India, Bolivia and Mali. *Journal of Rural Studies, 23*(1), 23–37. https://doi.org/10.1016/j.jrurstud.2006.02.006.

Roco, M. C. (2003). Broader societal issues of nanotechnology. *Journal of Nanoparticle Research, 5*(3–4), 181–189. https://doi.org/10.1023/A:1025548512438.

Roco, M. C. (2011). The long view of nanotechnology development: The National Nanotechnology Initiative at 10 years, pp. 427–445, https://doi.org/10.1007/s11051-010-0192-z.

Roco, M., & Bainbridge, W. S. (2001). *Societal Implication of Nanoscience and Nanotechnology*. Virginia: National Science Foundation.

Roco, M. C., Mirkin, C. A., & Hersam, M. C. (2011). Nanotechnology research directions for societal needs in 2020: Summary of international study. *Journal of Nanoparticle Research, 13*, 897–919. https://doi.org/10.1007/s11051-011-0275-5.

Rogers-Hayden, T., & Pidgeon, N. (2008). Developments in nanotechnology public engagement in the UK: "upstream" towards sustainability? *Journal of Cleaner Production, 16*(8–9), 1010–1013. https://doi.org/10.1016/j.jclepro.2007.04.013.

Roosen, J., Bieberstein, A., Blanchemanche, S., Goddard, E., Marette, S., & Vandermoere, F. (2015). Trust and willingness to pay for nanotechnology food. *Food Policy, 52*, 75–83. https://doi.org/10.1016/j.foodpol.2014.12.004.

Ross, M., Mcfarland, C., & Fletcher, G. J. O. (1981). The effect of attitude on the recall of personal histories. *Journal of Personality and Social Psychology, 40*(4), 627–634. https://doi.org/10.1037/0022-3514.40.4.627.

Rotberg, R. I. (2014). Good governance means performance and results. *Governance: An International Journal of Policy, Administration, and Institutions, 27*(3), 511–518. https://doi.org/10.1111/gove.12084.

Saidi, T. (2018). Perceived risks and benefits of nanomedicine: A case study of an anti-tuberculosis drug. *Global Health Innovation, 1*(1), 1–7. https://doi.org/10.15641/ghi.v1i1.496.

Sannino, D., Rizzo, L., & Vaiano, V. (2017). Progress in nanomaterials applications for water purification. *Nanotechnologies for Environmental Remediation*, 1–24. https://doi.org/10.1007/978-3-319-53162-5.

Schenk, M. F., Fischer, A. R. H., Frewer, L. J., Gilissen, L. J. W. J., Jacobsen, E., & Smulders, M. J. M. (2008). The influence of perceived benefits on acceptance of GM applications for allergy prevention. *Health, Risk and Society, 10*(3), 263–282. https://doi.org/10.1080/13698570802160947.

Scheufele, D. A., Corley, E. A., Shih, T., Dalrymple, K. E., & Ho, S. S. (2009). *Religious beliefs and public attitudes toward nanotechnology in Europe and the United States, 4*(February), 91–94. https://doi.org/10.1038/NNANO.2008.361.

Scheufele, D. A., & Lewenstein, B. V. (2005). The public and nanotechnology: How citizens make sense of emerging technologies. *Journal of Nanoparticle Research, 7*(6), 659–667. https://doi.org/10.1007/s11051-005-7526-2.

Schütz, H., & Wiedemann, P. M. (2008). Framing effects on risk perception of nanotechnology. *Public Understanding of Science, 17*, 369–379. https://doi.org/11.1077/0963662506071282.

Schwab, K. (2015). The fourth industrial revolution. *World Economic Forum, 10*(1), Switzerland. https://doi.org/10.1038/nnano.2015.286.

Sheila, D. (2017). Overview of an internationally integrated nanotechnology. *International Journal of Metrology and Quality Engineering, 8*, 8. https://doi.org/10.1051/ijmqe/2017002.

Siegrist, M. (2010). Predicting the future: Review of public perception studies of nanotechnology. *Human and Ecological Risk Assessment: An International Journal, 16*(4), 837–846. https://doi.org/10.1080/10807039.2010.501255.

Siegrist, M., Cousin, M. E., Kastenholz, H., & Wiek, A. (2007a). Public acceptance of nanotechnology foods and food packaging: The influence of affect and trust. *Appetite, 49*(2), 459–466. https://doi.org/10.1016/j.appet.2007.03.002.

Siegrist, M., Cvetkovich, G., & Roth, C. (2000). Salient value similarity, social trust, and risk/benefit perception salient. *Risk Analysis, 20*(3), 353–362. https://doi.org/10.1111/0272-4332.203034.

Siegrist, M., & Keller, C. (2011). Labeling of nanotechnology consumer products can influence risk and benefit perceptions. *Risk Analysis, 31*(11), 1762–1769. https://doi.org/10.1111/j.1539-6924.2011.01720.x.

Siegrist, M., Keller, C., Kastenholz, H., Frey, S., & Wiek, A. (2007b). Laypeople's and experts' perception of nanotechnology hazards. *Risk Analysis, 27*(1), 59–69. https://doi.org/10.1111/j.1539-6924.2006.00859.x.

Siegrist, M., Stampfli, N., Kastenholz, H., & Keller, C. (2008). Perceived risks and perceived benefits of different nanotechnology foods and nanotechnology food packaging. *Appetite, 51*(2), 283–290. https://doi.org/10.1016/j.appet.2008.02.020.

Simon, H. A. (1977). The logic of heuristic decision making. In *Models of discovery. And other topics in the methods of science*. D. Reidel Publisihing Company (pp. 154–175).

Sjöberg, L., Moen, B.-E., & Rundmo, T. (2004). Explaining risk perception, An evaluation of the psychometric paradigm in risk perception research. *Rotunde, Trondheim: Rotunde, 84*, 55–76. https://doi.org/10.1080/135753097348447.

Slovic, P., Finucane, M. L., Peters, E., & MacGregor, D. G. (2007). The affect heuristic. *European Journal of Operational Research, 177*(3), 1333–1352. https://doi.org/10.1016/j.ejor.2005.04.006.

Stanovich, K. E., & West, R. F. (2000). Individual differences in reasoning: Implications for the rationality debate? *Behavioral and Brain Sciences, 23*(5), 645–665.

Starr, C. (1969). Social benefit versus technological risk. What is our society willing to pay for safety? *Science, 165*, 1232–1238.

Stoker, G., Jennings, W., Evans, M., & Halupka, M. (2017). The impact of anti-politics on policymaking: Does lack of political trust matter? *UK political studies association conference* (pp. 1–26). 10–12 April 2017, Glasgow.

Tangau, W. M. (2017). Shaping the Malaysian Industry for the 4th Industrial Revolution, viewed 20 January 2018. https://www.akademisains.gov.my/download/YBMKEYNOTEADDRESS@SIAPCONFERENCE.pdf.

Tansey, J., & O'Riordan, T. (1999). Cultural theory and risk: A review. *Health, Risk and Society, 1*(1), 71–90. https://doi.org/10.1080/13698579908407008.

Tversky, A., & Kahneman, D. (1981). The framing of decisions and the psychology of choice. *Science, New Series, 211*(4481), 453–458. Viewed 20 June 2016. http://www.jstor.org/stable/1685855.

Tyshenko, M. G. (2014). Nanotechnology framing in the canadian national news media. *Technology in Society, 37*(1), 38–48. https://doi.org/10.1016/j.techsoc.2013.07.001.

UNESCAP. (2009). What is good governance? *United Nations Economic and social Comission for Asia and the Pacific*. https://doi.org/10.1016/B978-012397720-5.50034-7.

van Giesen, R. I., Fischer, A. R. H., & van Trijp, H. C. M. (2018). Changes in the influence of affect and cognition over time on consumer attitude formation toward nanotechnology: A longitudinal survey study. *Public Understanding of Science, 27*(2), 168–184. https://doi.org/10.1177/0963662516661292.

Vaughan, E., & Nordenstam, B. (1991). The perception of environmental risk among ethnically diverse groups. *Journal of Cross-Cultural Psychology, 22*(1), 26–60. https://doi.org/10.1177/0002764213490695.

Vishwakarma, V., Samal, S. S., & Manoharan, N. (2010). Safety and Risk Associated with nanoparticles—A review. *Journal of Minerals and Materials Characterization and Engineering, 9*(5), 455–459.

West, G. H., Lippy, B. E., Cooper, M. R., Marsick, D., Burrelli, L. G., Griffin, K. N., & Segrave, A. M. (2016). Toward responsible development and effective risk management of nano-enabled products in the U.S. construction industry. *Journal of Nanoparticle Research, 18*(2), 1–27. https://doi.org/10.1007/s11051-016-3352-y.

Wiek, A., Foley, R. W., & Guston, D. H. (2012). Nanotechnology for sustainability: what does nanotechnology offer to address complex sustainability problems? *Nanotechnology for Sustainable Development* (pp. 371–390). New York: Springer, Cham.

Wildavsky, A. (1987). Choosing preferences by constructing institutions: A cultural theory of preference formation. *The American Political Science Review, 81*(1), 3–21. https://doi.org/10.2307/1960776.

Zainal Abidin, I. S. (2018). January 26. Embracing the fourth industrial revolution. *New Straits Times*, viewed 25 February 2018. https://www.nst.com.my/opinion/columnists/2018/01/328868/embracing-fourth-industrial-revolution.

Zhang, J., Wang, G., & Lin, D. (2015). High support for nanotechnology in China: A case study in Dalian. *Science and Public Policy, 43*(1), 115–127. https://doi.org/10.1093/scipol/scv020.

Zimmer, R., Hertel, R., Böp. G. -F., & Hertel, R. (2010). *Risk Perception of nanotechnology—analysis of media coverage*, Berlin: BfR Wissenschaft, viewed 16 June 2016. http://www.bfr.bund.de/cm/350/risk_perception_of_nanotechnology_analysis_of_media_coverage.pdf.

# Integrating Responsible Care Through Quality, Environmental, Health and Safety Management System for Chemical Industries in Malaysia

Nur Khairlida Muhamad Khair, Khai Ern Lee, Mazlin Mokhtar, and Choo Ta Goh

**Abstract**

The development of chemical industries in Malaysia has raised the need to strengthen the chemicals management to avoid incidents which could put the environment, health and safety at risks. Responsible Care program was launched in 1994 to promote sustainable management of chemicals throughout the product chain as a commitment to build trust in the chemical industries. Although Responsible Care program was developed to restore public's trust toward chemical industries, there are issues debated particularly on the effectiveness of its self-regulatory program whether could achieve the objective of practicing sustainable chemical industry management. In this chapter, a concept of integrating Responsible Care through quality, environmental, health and safety management system for chemical industries is laid out, taking Malaysian chemical industries as a case study. Considering the challenges faced by the signatory chemical companies, a simplified and integrated management system for Responsible Care has been developed by incorporating ISO9001, ISO14001 and OHSAS18001 with Responsible Care program. The simplified and integrated management system for Responsible Care is expected to assist chemical companies in enhancing the implementation of Responsible Care program and promoting more participation of chemical companies in Responsible Care program for the sustainability of chemical industries.

**Keywords**

Responsible care • Quality • Environment • Health • Safety • Chemical industries • Sustainability

## 1 Chemical Industries

Malaysia's economy grows rapidly since its independence in 1957. At the earlier years of independence, the country's economic activities relied heavily on rubber and tin commodities which accounted for 70 percent of the total exports (Economic Planning Unit 2013). However, in the 1980s, Malaysia has transformed into one of the largest exporters among the developing countries whereby it attracted foreign direct investment (FDI) into the electronics industry sector (Athukorala and Menon 1996). Malaysia's strategy in advancing the industrial sector in the mid-1980s begun with the launch of the National Industrial Policy and Industrial Masterplan, which aim to make Malaysia a nation that focuses on higher value-added economic activities, as well as to reduce dependence on upstream commodities, such as tin and rubber (Central Bank of Malaysia 2013). As a result of the introduction of the policies, the Malaysian industrial sector has grown rapidly and contributed 36.8 percent to gross domestic product (GDP) in 2016 compared with 17.2 percent in 1980.

Malaysian industry sector can be classified into resource-based industries or those involved with downstream processing and manufacturing of products that use the country's major resources. Examples of resource-based industries include manufacturing of food, beverages, tobacco, timber, chemicals and chemical products, rubber products, non-metallic products and non-ferrous metals. Non-resource-based industries comprise clothing and textiles, steel and iron, metal products, electronic equipment and transport equipment. According to May (2000), the manufacturing sector particularly resource-based chemical industry has played an important role in sustaining economic growth for a long time. In 1998, the resource-based industry contributed 51% to the total national production revenue and industry with the highest average annual growth is non-ferrous metal industry and chemical industry. The annual growth of chemical industries in the manufacturing sector has been a

N. K. M. Khair · K. E. Lee (✉) · M. Mokhtar · C. T. Goh
Institute for Environment and Development (LESTARI), Universiti Kebangsaan Malaysia, Bangi, 43600, Selangor, Malaysia
e-mail: khaiernlee@ukm.edu.my





K. E. Lee (ed.), *Concepts and Approaches for Sustainability Management*, Advances in Science, Technology & Innovation, https://doi.org/10.1007/978-3-030-34568-6_2

catalyst to the Malaysian economy in which Malaysian chemical industries continue to grow rapidly, contributing 6 percent to the manufacturing exports in Malaysia (Department of Statistics Malaysia 2011).

According to Lee et al. (2015), chemical industries in Malaysia can be categorized into eight sub-sectors, i.e., oleochemicals, industrial gases, petrochemicals, pesticides, fertilizers, paints, coating fluids and biodiesel. In 2015, the Federation of Malaysian Manufacturers (FMM) has recorded 535 chemical companies and chemical-related companies throughout Malaysia (Federation of Malaysian Manufacturers 2015). These companies are domestic companies as well as multinational companies that produce chemicals and also handle chemicals (Khair et al. 2017). Most of the major chemical products in Malaysia are produced by giant companies, such as Petronas, CCM and so forth. However, small and medium companies also produce chemical products, such as chemical, pharmaceutical, plastic and rubber products (MITI 2010). The oleochemical sector of Malaysia is one of the largest sectors in the world by supplying 20 percent to the global capacity, contributing 21.9 percent to the total chemical exports (MATRADE 2014).

The rapid development of the chemical industries in Malaysia has now triggered the need to strengthen the management of chemicals in order to prevent incidents involving chemicals that could threaten human health and safety as well as the environment. Chemical industries improve people's quality of life and contribute to the country's economic development. However, chemical industries could also harm the environment, human health and safety if an incident occurs due to the irresponsible attitudes and behaviors of chemical industries toward the society and its environment. In 1984, a disastrous chemical incident killed more than 2500 lives, and more than 200,000 people in Bhopal, India, were exposed to 30 tons of methyl isocyanate (MIC) toxic gas released by the Union Carbide pesticides (Dhara and Dhara 2002). Following the incident, the Canadian Chemical Producers' Association (CCPA) launched Responsible Care program (Fig. 1) in 1985, aiming to improve the performance of chemical industries in environmental, health and safety aspects (Belanger et al. 2009b). Soon after the launch in Canada, the program has also been introduced in several other countries (Gunningham 1995).

**Fig. 1** Logo of Responsible Care program

## 2 Responsible Care Program

Responsible Care program was first implemented in Canada since 1985, after Bhopal incident. The main focus of Responsible Care program was to restore the image of chemical industries and regain public confidence in chemical industries (Givel 2007; Gunningham 1995). In addition, Responsible Care program is seen as part of the commitment and responsibility of chemical industries to contribute to the community as well as to reduce any adverse impacts on the environment and society (Belanger et al. 2009a; Moffet et al. 2004.

Implementation of Responsible Care program has evolved globally with the same purpose, in response to the community's concerns in the environmental, health and safety aspects of chemical industries. The involvement of chemical companies in the program is subjected to the organization of chemical industries in their respective countries. As in the USA, the Chemical Manufacturers' Association (CMA) launched Responsible Care program in 1988 and then obliged all members of the association to implement Responsible Care program as one of the conditions of their memberships in the association (King and Lenox 2000).

As shown in Fig. 2, Responsible Care program has been widely implemented in OECD countries and is currently implemented by developing countries, including South Africa and Asia countries (Tapper 1997). The implementation of Responsible Care program is subjected to the national chemical industries association in each country to determine the specific ways, principles and priorities that should be addressed. Every company implementing Responsible Care program will be assisted by local chemical association by providing them implementation framework (Gunningham 1995; Lee et al. 2015; Tapper 1997. Companies involved in implementing Responsible Care program should commit to ensure that all parties, include company employees, customers, distributors, suppliers and communities working together, to ensure the success of Responsible Care program.

### 2.1 America and Europe Continents

Since the launch of Responsible Care program in 1985, many countries have adopted this initiative into their country's chemical industries. The implementation of Responsible Care program varies for each country in which it belongs. Responsible Care program was launched in the USA since 1988 by the American Chemical Association (ACC). Throughout the implementation of the program, three codes of management practices have been adopted, which are product safety, process and security codes (American Chemical Council 2015).

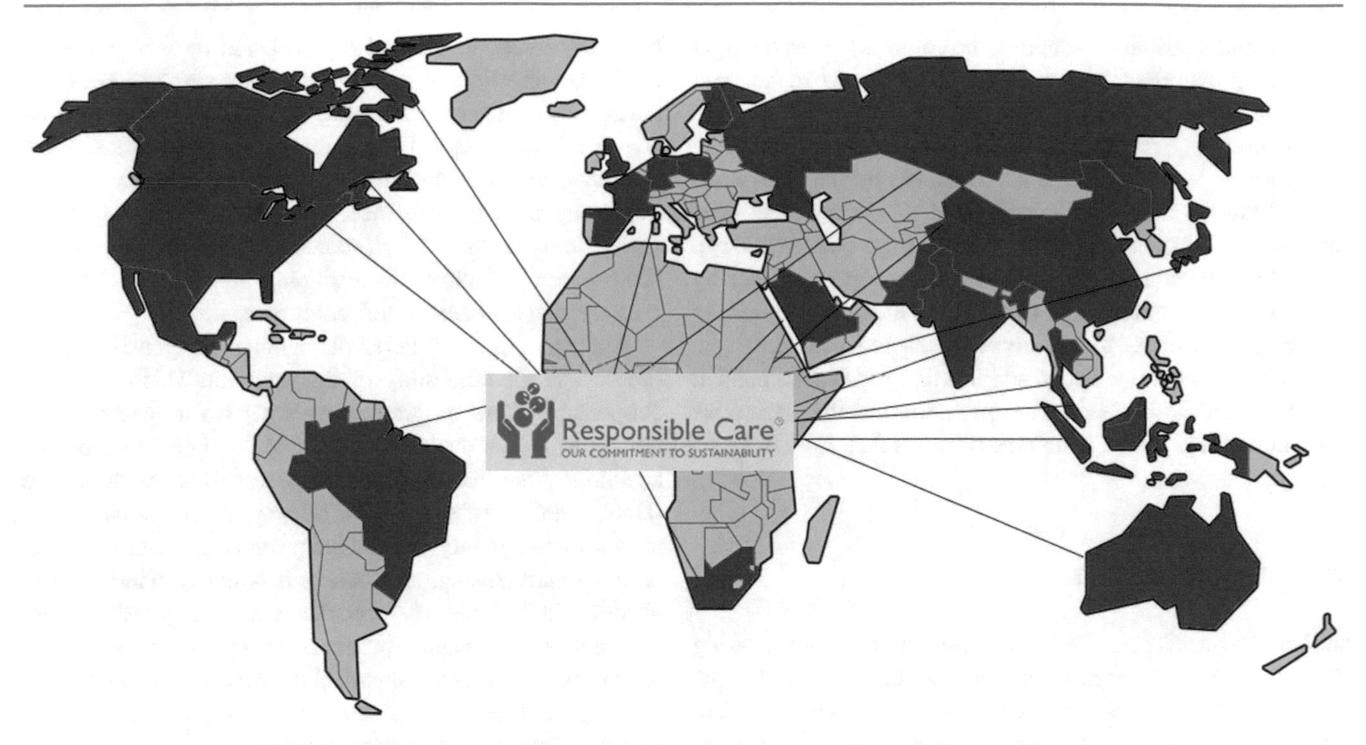


**Fig. 2** Countries that implement Responsible Care program

The initiative of the USA in promoting chemical industries toward sustainability was followed by chemical industries in the Europe continent. The development of Responsible Care program in the Europe continent began after this initiative has been developed in the USA. The implementation of Responsible Care program in Europe is regulated by the Council of European Chemical Industry (CEFIC), comprising European countries, namely Germany, Greece, France, Russia, UK, Finland and others. In 2010, CEFIC and its council members adopted European Responsible Care security code (CEFIC 2016), and CEFIC is committed to promote and ensure the implementation of Responsible Care program in Europe goes on consistently. CEFIC is also responsible for developing and ensuring that every council member is able to implement Responsible Care program into their management system. At the same time, CEFIC also actively participates in Responsible Care program in the global level, such as becoming a member of International Council of Chemical Association (ICCA) (CEFIC 2016).

In addition to the USA, taking the implementation of Responsible Care program in Canada as an example, they have employed three codes of management practice, comprising operating, product stewardship and accountability codes. Operating code includes managing their equipment and facilities to ensure that they operate in a safe and responsible manner. As such, the company should strive to achieve continual improvement in environmental performance and reduce resource consumption. For product stewardship code, companies need to conduct regular survey on the impact and safety of the products they produce, including any services and technologies used. Besides ensuring product safety, companies are also encouraged to have good communication with each stakeholder involved in the overall product manufacturing process. For accountability code, companies are required to communicate with communities living close to the company or community along the transportation corridor as well as for other stakeholders involved in the manufacturing process (Chemical Industry Association of Canada 2015).

## 2.2 Australia Continent

Besides America and Europe continents, Responsible Care program has also been introduced in Australia since 1989 and stewarded by the Australian Chemical Industry Organization (ACIC) which is now converted to the Plastics and Chemical Industry Association (PACIA). At the beginning of its launch, Responsible Care program has eight codes of management practices, consisting distribution, community awareness and emergency response, waste management, warehousing and storage, the community has the right to know, product stewardship, manufacturing, as well as research and development (Gunningham 1995). Throughout the implementation of Responsible Care program, there are amendments and changes have been made in which PACIA has implemented six codes of management practices only. The six codes of management practices comprise employees' safety and health, storage and distribution security, the community has the right to know, manufacturing safety, product stewardship and environmental maintenance (PACIA 2015).

Meanwhile, the implementation of Responsible Care program in New Zealand is administered by the New Zealand Chemical Industry Council (NZCIC) or has now been changed to Responsible Care New Zealand (NZCIC 2016). They have launched Responsible Care program in 1991 by promoting the commitment of chemical industries to operate safely and harmless to the environment and human beings. NZCIC also offers advisory and training services to council members on Responsible Care program and the Hazardous Substances and New Organisms (HSNO) Act 1996 (NZCIC 2016).

## 2.3 Asia Continent

Asian countries are not left behind in implementing Responsible Care program in their chemical industries. The organization involved in implementing Responsible Care program in Asia is the Asia Pacific Responsible Care Organization (APRO). Among the countries participated in APRO are Japan, Taiwan, China, Philippines, Thailand, Malaysia, Singapore, Indonesia, India, Australia and New Zealand (Shiozaki 2009). The APRO is under the supervision of the International Council of Chemical Association (ICCA), but only focuses on matters related to Responsible Care at the Asia Pacific level. In 1995, APRO held its first conference, the Asia Pacific Responsible Care Conference (APRCC) in Hong Kong (Shiozaki 2009). The implementation of Responsible Care program in Asia adopts six codes of management practices, namely distribution, community awareness and emergency response, pollution prevention, process safety, employee health and safety and product stewardship (CICM 2015; JCIA 2011; RCI 2015; SCIC 2015). The aforementioned Asian countries are also actively involved in membership in the Responsible Care Leadership Group (RCLG) and also focus their role on Responsible Care Global Charter, Responsible Care development, performance improvement and reporting, as well as communication with stakeholders (Shiozaki 2009). For example, Chemical Industries Council of Malaysia (CICM) is actively involved in sharing experiences with international chemical associations as well as members in RCLG and APRO (CICM 2015).

## 2.4 Issues and Challenges of Responsible Care Program

Although Responsible Care program was developed to restore the public's trust toward chemical industries, there are many issues that are always debated by researchers on the implementation of this program. Among the issues, the effectiveness of this self-regulatory program is of particular concern. While this self-regulatory program is implemented by chemical industries' own association with the aim of improving their performance, the question of its effectiveness is raised whether Responsible Care program could achieve the objective of practicing sustainable chemical industry management is still controversial.

Another criticism in implementing Responsible Care program is the credibility of the program. This is because Responsible Care program does not have a specific audit system to assess the performance of its signatories, but its signatories are required to self-assess their performance (Hook 1996). Self-assessment of program performance has shown that signatory companies promoting themselves like corporate advertising, and hence, it is not convincing. The situation is increasingly complex when Responsible Care program is a voluntary program, as the public and environmental scientists are skeptical about its effectiveness. It is because for them such a self-regulatory program is just a greenwashing that companies demonstrate their commitment to sustainability by participating in the program, but in fact they fail to minimize environmental impact (Lyon and Maxwell 2011). In addition, a survey was conducted by Grolleau et al. (2007) on 86 chemical companies in France which shows that chemical company shareholders are keener to implement ISO14001 than Responsible Care program due to its lack of credibility.

In addition, Glachant 2007 argues that companies participating in this self-regulatory program will not necessarily reduce their pollution. This is because there are companies that use the strategy to participate in this self-regulatory program without any intention to comply with it, simply to avoid tighter legislative intervention on them (Maxwell et al. 2000). In fact, there are companies that seem to fulfill their commitments to the programs that they are participating but not directly addressing environmental issues (Calcott 2010). Morgenstern and Pizer 2007 also note that companies involved in self-regulatory or voluntary programs do not show significant reduction of pollution. Similarly, Gamper-Rabindran 2006 argues that self-regulatory programs lead to the increase of more emissions of toxic substances into the environment compared to companies that are not participating in the programs. The percentage of total pollution reduction in companies involved in self-regulatory is also low.

In addition to the weaknesses in the credibility of Responsible Care program, the problem of such self-regulatory programs is the lack of a system to monitor the company's compliance with the program and impose action on non-compliance companies (King and Lenox 2000). Programs conducted without a monitoring system and actions for non-compliance will result in failure to take action against companies that violate program codes (Lenox and Nash 2003). This will cause a free ride problem whereby the program implementation is only on paper but not properly implemented. However, recommendations for using independent third parties in Responsible Care program assessment have been implemented, but this assessment is not implemented by all companies involved (Prakash 2000).

The next issue raised about Responsible Care program is that most companies only report their success in reducing environmental pollution but never report any of their failures (Tschopp 2005). This will give an incomplete picture to the public because of a non-transparent report. However, Responsible Care code, namely community awareness and emergency response, supports the concept of "people have the right to know" the activities of chemical companies (Gunningham 1995). Companies joining Responsible Care program should be clearer about their performance in each published report so no one will question such issue.

In the context of pollution control, King and Lenox 2000 found that companies participating in Responsible Care program reduce their pollution at a slower pace than non-participants. Additionally, some researchers have discussed in their studies that companies that implement their own regulatory programs do not necessarily reduce pollution, but even further increase pollution (Gamper-Rabindran and Finger 2013; Morgenstern and Pizer 2007). This indicates that though Responsible Care program has pollution prevention code, the evidence shown does not reflect the reality.

Similar to other program implementations, Responsible Care program also has some challenges that chemical companies need to face in implementing this program into their company. The implementation of Responsible Care program in small and medium companies faces problems, such as lack of funds, labor and expertise to implement Responsible Care program (Gunningham 1995). In addition, similar challenges faced by small and medium chemical companies in Malaysia to implement Responsible Care programs, such as lack of labor, good infrastructure, trained experts and individuals, weak program enforcement and over-dependency on foreign workers (Lee et al. 2015; Rampal and Nizam 2006).

Different companies need different reporting for Responsible Care program on top of other management systems. This leads companies to spend more time in producing report rather than implementing the program (Howard et al. 1999). In addition, the code of behavioral management of Responsible Care program requires specialized training to be implemented into the company. However, little or no training on Responsible Care program is practically given to employees. Most of the exercises carried out are informal and do not specifically address Responsible Care program (Howard et al. 1999).

A study conducted on chemical companies in Greece found that the challenges they faced to implement Responsible Care program are less knowledge about the program (Evangelinos et al. 2010). In addition, the high level of bureaucracy to enable companies to implement Responsible Care program, the low recognition status of Responsible Care program, conflicts with other management systems, excessive program requirements, requiring financial support from the government, low level of environmental awareness in chemical sector and low perspective benefits (Evangelinos et al. 2010). Additionally, challenges such as low level of Responsible Care program recognition and high program implementation costs are also identified by Prakash (1999) for the chemical industries in the USA.

## 3 Case Study: Responsible Care in Malaysian Chemical Industries

Malaysia is not left behind to implement this program where the Chemical Industries Council of Malaysian (CICM) launched Responsible Care program in 1994 (CICM 2015). Responsible Care program is implemented on the basis of sustainable development and the use of chemicals as well as communicating directly with stakeholders (Lee et al. 2015).

The implementation of Responsible Care program in Malaysia has lasted for over 20 years. Throughout its implementation, CICM is actively involved in various activities involving international or local Responsible Care programs. For example, CICM is one of the members of the Responsible Care Leadership Group (RCLG) in the International Chemical Society Committee (ICCA) which is the governing body for the implementation of Responsible Care program internationally and participates in the Asia-Pacific Responsible Care Organization (APRO) (CICM 2015). CICM's active involvement in various programs concerning Responsible Care shows that CICM is committed to ensuring that chemical industries in Malaysia are not left behind the current development of global chemical industries. However, the challenge in ensuring the continued implementation of Responsible Care program in Malaysian chemical industries is still inevitable. Therefore, improvements in chemical industry management system are needed to achieve the goal of making chemical industries in the country shift toward sustainability.

Seven codes of management practices were launched in stages by CICM from 1995 to 1999 as shown in Table 1. Responsible Care program is implemented in accordance to guiding principles as well as the seven codes of management practices, consisting community awareness and emergency response, distribution, pollution prevention, process safety, safety and health of employees, product stewardship and security. All these codes of management practices are

**Table 1** Responsible Care codes of management practices and purposes

| | Codes of management practices | Date launched | Purposes |
|---|---|---|---|
| 1 | Distribution code | August 3, 1995 | To reduce the potential for harm posed by the distribution of chemical to the general public, employees and the environment |
| 2 | Community awareness and emergency response code | November 2, 1996 | To work with nearby communities to understand their concerns and to plan and practice for emergencies |
| 3 | Pollution prevention code | November 25, 1997 | To achieve ongoing reduction in the amount of all pollutants released into the environment |
| 4 | Process safety code | May 19, 1999 | To prevent fires, explosions and accidental chemical releases |
| 5 | Employee health and safety code | November 17, 1999 | To protect and promote the health and safety of people working at or visiting company sites |
| 6 | Product stewardship code | November 17, 1999 | To make health, safety and environmental protection a priority in all stages of product's life, from design to disposal |
| 7 | Security code | December 12, 2017 | To help protect people, property, products, processes, information systems by enhancing security, including security against potential terrorist attack, throughout the chemical industry value chain |

*Source* Responsible codes of management practices manual (CICM 1999, 2019)

developed to meet the needs of the community, the current and future importance of chemical products and to support the development of law enforcement, chemical companies' activities and protecting the environment (Hook 1996). According to the American Chemical Council 1995, codes of management practices are the mechanisms in adopting Responsible Care program, the adoption of these codes of management practices into management system makes the health, safety and environmental performance better and the openness of chemical companies to the public becomes more transparent.

## 4 Adoption and Implementation of Responsible Care in Chemical Industries

### 4.1 Approach

The adoption and implementation of Responsible Care program in chemical industries have received attentions in many respects, especially the issue of Responsible Care program that is implemented without any regulation that may apply to companies in the event of a violation of the program code. In addition, the credibility of Responsible Care program is also compared with other management systems, such as ISO standards which have a more effective impact than Responsible Care. As such, the American Chemical Council (ACC) has developed a new approach by integrating Responsible Care program into ISO14001 management system as RC14001. In the process, Responsible Care program has been revised and incorporated into Deming's principles that include "Plan-Do-Check-Act" processes (Fig. 3). In 2002, there were major changes in the four key areas of codes of management practices, third-party verification, performance data and security requirements (Yohonn 2004). However, after major changes in Responsible Care program, chemical companies have not implemented all six codes of management practices; they are required to implement a new system of Responsible Care Management System (RCMS) into their company.

This RC14001 management system acts as a tool in combining the environment, health and safety aspects into business operations to improve the performance of the company in addition to meeting commercial and stakeholders' needs. This new approach also includes the specific elements of Responsible Care program which is more focused on transport, supply chain management, product control and stakeholder engagement (Yohonn 2004). Companies interested in RC14001 certification need to first meet the ISO14001 certification because this RC14001 uses the full ISO14001 guideline.

In addition, the policy used in RC14001 requires individuals to manifest responsibility from the management, open communication with stakeholders and commit to the principles of Responsible Care program (Phillips 2006). Since RC14001 is the result of the integration of two management systems into an integrated management system, the scope of ISO14001 is extended to include Responsible Care elements. Therefore, Responsible Care and health, safety and security are included in every environmental aspect contained in ISO14001 management system (Yohonn 2004).

In an analysis of RC14001 management system conducted by Yohonn 2004, different types of indicators are categorized to assess the company's performance. Among them are the driver of implementation, the impact of employee engagement, the overall effect of management

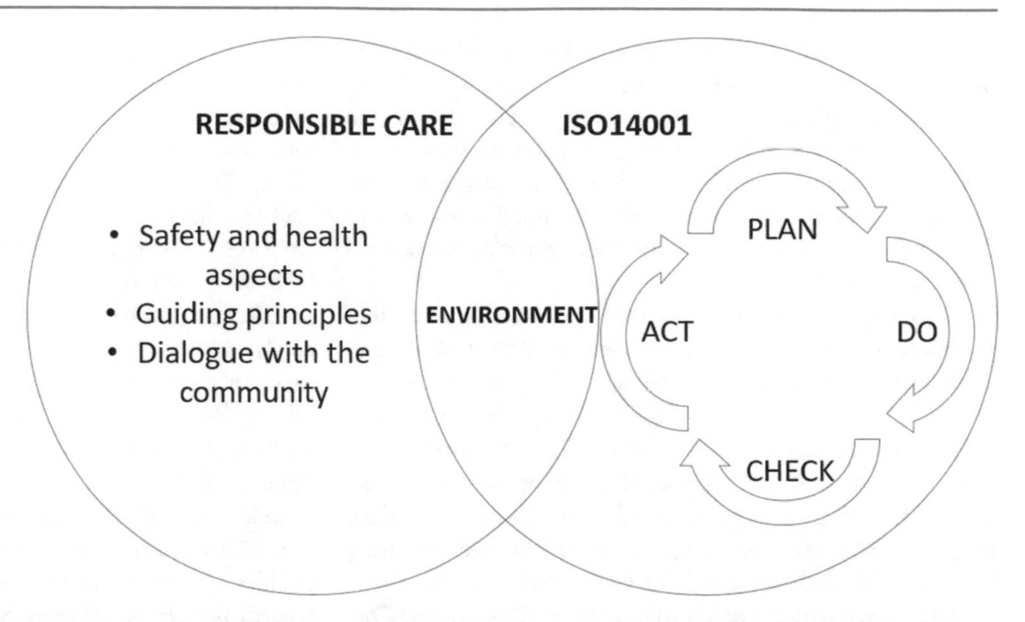


**Fig. 3** Responsible Care and ISO14001 (Yohonn 2004)

system, the effect of management commitment, the understanding of relevant rules, the effect of RC14001 on EHS performance and the enhancement of EHS management. The results of the analysis show that RC14001 has a positive impact on employee involvement and overall management system. However, RC14001 management system focuses only on the environmental aspects, which is not comprehensive. A non-exhaustive approach of RC14001 management system requires a new, more holistic, comprehensive system that covers all the elements in Responsible Care program. Considering Responsible Care program is not focusing on the environment per se, other elements, such as product safety, employee health, community awareness and process security should be included.

## 4.2 Concept

Although Responsible Care program has been launched for almost 30 years, the program is still receiving criticisms about its credibility and the community is also skeptical about the program by saying that Responsible Care program is one of the industry's strategies to improve their image (Castleman 1997). In recent years, sustainable development of chemical industries has become an important agenda globally. There are many types of management systems that have been used to help the industry to create sustainable competitiveness. Among the common management systems used in chemical industries are ISO9001 (quality management), ISO14001 (environmental management), OHSAS18001 (safety and health management) and Responsible Care program. While all these management systems are used for various objectives and targets, it is clear that each management system shares the same techniques and principles.

The study conducted by Delmas and Montiel (2008) on the implementation of management systems, such as Responsible Care, ISO9001 and ISO14001 in chemical industries shows that those management systems are complementing rather than competing among each other. Additionally, they also argue that environmental should not be managed separately but should be complementing with other management systems. This clearly demonstrates that integrating all standards into a simplified management system that is consistent with the organization's objectives and strategies is better than having multiple management systems at one time. Therefore, the main purpose of integrating Responsible Care program into quality, environmental, health and safety management system is to strengthen the credibility of Responsible Care program implemented in chemical companies.

There are many studies that discuss the advantages of implementing an integrated management system. Among them is to promote collaboration between management systems and cost savings (Simon et al. 2013). In addition, the implementation of this integrated management system is also related to the factors that cause chemical companies to embrace it in their companies' management system. The implementation of an integrated management system demonstrates many benefits to the companies. However, there are difficulties and challenges that must be addressed by an organization. Several studies on integrated management system have discussed barriers to implement an integrated management system, consisting of internal barriers as well as external barriers. Internal barriers comprise problems arising in companies, such as lack of human resources, lack of company staff motivation, acceptance and recognition of new systems by staff, organizational structures and working culture within the company (Oskarsson and von Malmborg

2005; Simon et al. 2013; Zeng et al. 2006). External barriers are the constraints arising out of the company, including technical guidance, verification bodies, stakeholders and customers, as well as institutional environment (Zeng et al. 2006). Although the implementation of an integrated management system has constraints in some respects, its implementation into the companies provides substantial benefits for the company.

A conceptual framework has been developed as a result of literature review on important documents pertaining to relevant standards and Responsible Care program whereby it comprises ISO9001, ISO14001, OHSAS18001 and Responsible Care program (Khair et al. 2018). The identified aspects from each standard will be used as inputs for conceptual framework to guide the integration of Responsible Care program into an integrated management system using "Plan-Do-Check-Act" principles as shown in Fig. 4.

There are two approaches in integrating Responsible Care program into an integrated management system whereby a top-down approach is to simplify the integration process between management systems (Asif et al. 2013). In addition to using top-down approach, this approach also uses a meta-management approach or system approach (Asif et al. 2013; Jonker and Karapetrovic 2004; Karapetrovic and Jonker 2003). Referring to Sabatier (1986), the top-down approach has important features, i.e., the approach begins with the policy decision by the supreme party. The policy decision will be the guide or the basis for an organization to act in line with the design.

In this context, a top-down approach focuses on integrating Responsible Care program codes into quality, environmental, health and safety management system, aiming to improve the credibility of Responsible Care program in chemical industries. This integration also provides a systematic approach in forming Responsible Care objectives. In addition, this approach enables companies to identify and control the effects of their company activities, products or services as well as to improve their performance in quality,

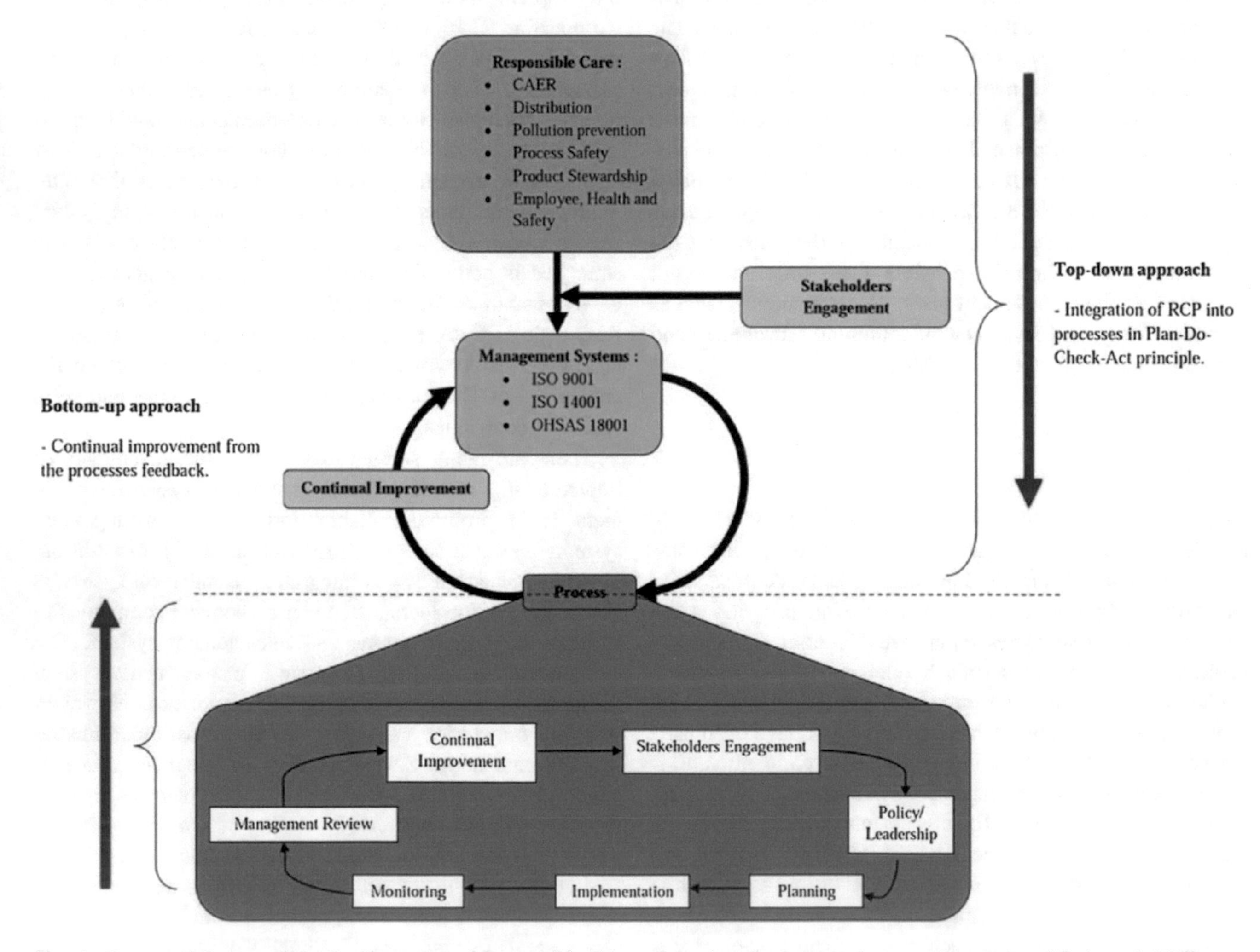


**Fig. 4** Conceptual framework for the integration of Responsible Care program into the integrated management system (Khair et al. 2018)

environment, health and safety. Six codes of management practices, such as community awareness and emergency response, distribution, pollution prevention, process safety, safety and health of employees and product stewardship are integrated using the Plan-Do-Check-Act (PDCA) process cycle into ISO9001, ISO14001 and OHSAS18001 management system. Among the processes involved are policies and leadership, planning, implementation, monitoring, management reviews and continuous improvement (Khair et al. 2018). The top-down approach focuses on stakeholder needs, such as employees, contractors, suppliers, distributors and product recipients involved in the implementation process of Responsible Care program. This approach also introduces the necessary structures at all levels of the organization, defining the needs of Responsible Care program and making communication and information delivery function more systematically.

According to Asif et al. (2013), they have referred the bottom-up approach as a guideline for community-related development focusing on stakeholder's engagement among communities and the development of indicators related to community needs, such as corporate roles in providing a good quality of life. They also argue that companies need this bottom-up approach to connect and foster good relationships between companies and communities. In addition, companies also need to interact with less powerful stakeholders, such as local communities. This is for them to better understand how their business operations have affected the local communities as well as improving the standard of living of the communities.

The use of the bottom-up approach in the conceptual framework involves continuous process of feedback generated from all the processes and stakeholders. This is to ensure that all stakeholders have a good quality of life. In addition, this bottom-up approach also encourages companies to interact with the communities as most companies only interact with stakeholders directly involved in the company's operations, while interactions with the communities, local authorities and nearby companies are often forgotten. Therefore, this bottom-up approach is appropriate for the implementation of Responsible Care program because it is in line with the program's objective to foster a good relationship between the company and the communities.

## 5 Integration of Responsible Care into QEHS Management System

There are seven major processes, and 28 aspects have been identified in Responsible Care program and ISO9001, ISO14001 and OHSAS18001 that could be integrated into Responsible Care, Quality, Environmental, Health and Safety Management System. The overall processes and aspects are shown in Table 2.

### 5.1 Processes

(i) Policy and Leadership

The policy is referred to as the basis or guide used in a company to determine actions in decision-making processes related to the management of chemical company. According to Yukl (1989), it is defined in terms of individual traits, leader behavior, interaction patterns, relationship roles, followers' perceptions, influence on followers, influence on task goals and influence on organizational culture. In the context of this conceptual framework, leadership refers to management actions in influencing employee participation in every process involved in the company's activities.

**Table 2** Processes and aspects for integrating Responsible Care into QEHS Management System

| Processes | Aspects |
|---|---|
| Policy and leadership | Policy, participation |
| Planning | Emergency response planning, disaster plan, design, risk control, waste, communication, documentation |
| Implementation | Training, communication, documentation, selection, emergency preparedness |
| Monitoring | Assessment and monitoring, record, preventive and corrective actions |
| Management review | Frequency, communication effectiveness, safety, waste, feedback, design review |
| Continual improvement | Learning process, benchmarking, system gap |
| Stakeholders engagement | Stakeholders |

(ii) Planning

Planning is to create objectives and processes needed to deliver results according to expectations. Planning for each code of management practices of Responsible Care program is required so that the policy can be implemented effectively. Each code has different objectives to achieve. As such, the signatory company of Responsible Care program needs to have appropriate planning in implementing codes of management practices.

(iii) Implementation

Implementation is to implement the plans that have been made. In addition, implementation also involves the process of data collection for the purposes of charts and analyses in the phases of check and act. The implementation of Responsible Care program codes of management practices covers six sub-processes, namely (i) documentation, (ii) selection, (iii) training, (iv) communication, (v) outreach and (vi) emergency preparedness. Documentation involves the process of documenting any procedures, information or activities undertaken in implementing Responsible Care program. For the selection of contractors, manufacturers, suppliers, distributors and containers, company needs to select and place some basic criteria in the selection. In addition, company is required to provide training that is appropriate to the objectives and goals that have been designed. At the same time, company is required to ensure the communication involved in each process is carried out properly. The implementation of outreach is the involvement of company with outsiders, such as community, government and media. Lastly, company is required to ensure that emergency preparedness measures are well implemented.

(iv) Monitoring

Monitoring is conducted to review the outcome of any activities implemented. The results will be compared with the expected outcomes of the goals outlined in the planning stage to identify any discrepancies that will be needed for management review. Monitoring for each code of management practices in Responsible Care program includes (i) evaluation and monitoring, (ii) records and (iii) corrective and preventative actions. Evaluation and monitoring involve risk assessment, performance advancement, community participation, incident investigation, communication, vehicle inspection and distributor qualification in distribution. Records include records of hazardous chemicals, maintenance and inspection tools, waste management, safety data sheet in product control and distribution process, employee health and safety programs and exposure assessment.

(v) Management Review

Management review aims to review all the processes, improve goals, revise methodology or develop new plans that are more appropriate based on the outcomes obtained from all the processes. Management review covers (i) frequency, (ii) communication effectiveness, (iii) security, (iv) waste, (v) feedback and (vi) design revision. Company is required to identify the frequency of review on each code of management practices. For communication effectiveness, signatory company should review the performance of communication involved in the codes of management practices. Security involves a review of the identified risk characteristics, while waste involves a review of the company's waste management performance. Subsequently, feedback involves the review of feedback from distributors. Design revision involves the process of checking the design of tools and plants that are potentially risky.

(vi) Continual Improvement

Continual improvement requires the company to improve its performance by removing irrelevant activities, improving efficiency, reducing costs and improving process capabilities. It also involves repeated processes aiming to improve Responsible Care program performance as a whole in line with the company's policies and goals, such as (i) learning process, (ii) benchmarking, (iii) gaps between management systems and iv) way forward.

(vii) Stakeholders Engagement

Stakeholders are those involved directly or indirectly in the management processes, consisting employees, contractors, suppliers, distributors, local communities, other nearby companies and local authorities. Stakeholders such as employees, contractors, suppliers and distributors are principally involved in the management processes where all the stakeholders should work together in ensuring the effectiveness of the management system implemented. Unlike stakeholders such as local communities, other nearby companies and local authorities who are only involved in a continuous process of feedback where their feedback is used to improve company's performance and help company to understand their needs.

## 5.2 Integrated Management System for Responsible Care

A simplified and integrated management system for Responsible Care has been developed whereby it integrates six Responsible Care codes of management practices, namely the community awareness and emergency response, process safety, pollution prevention, product stewardship, distribution and employee safety and health codes into the management system that encompasses ISO9001, ISO14001 and OHSAS18001. Using the process approach, "Plan-Do-Check-Act," the simplified and integrated management system for Responsible Care covers seven processes, i.e., policy and leadership, planning, implementation, monitoring, management review, continual improvement and stakeholder engagement as shown in Fig. 5. The simplified and integrated management system for Responsible Care is developed to fill the gaps of RC14001 management system, as this management system encompasses environmental, health and safety aspects to form a simpler and holistic implementation process (Khair et al. 2018, 2019).

### 5.2.1 Policy and Leadership

Policy and leadership are the first process that shall be undertaken by signatory company whereby the policy set by the leadership of signatory company will be used as a guideline to define actions in decision-making processes. The policy adopted should meet the objectives and goals of the simplified and integrated management system for Responsible Care. Whereas, leadership plays a role in influencing employees' participation in every process of the signatory company, and it can be demonstrated through the commitment by the top management through actively engaging in company activities and communicating with all levels of staff. Table 3 shows the management practices involved in the policy and leadership process of the simplified and integrated management system for Responsible Care.

### 5.2.2 Planning

Planning is the second process that should be undertaken by signatory company whereby signatory company is required to determine the objectives and goals of the simplified and integrated management system for Responsible Care. Table 4 shows the management practices involved in the planning process of the simplified and integrated management system for Responsible Care. Through planning, signatory company must ensure that the plans designed are appropriate and follow code of management practices implemented within the company.

### 5.2.3 Implementation

Upon identifying the goals and objectives to be achieved at the planning stage, the implementation is the stage of implementing the plans that have been made. Table 5 shows the management practices involved in the process of implementation and sub-processes of the simplified and integrated management system for Responsible Care. The six sub-processes include (i) documentation, (ii) selection of contractors, manufacturers, suppliers, distributors and containers for delivery purposes, (iii) training, (iv) communication, (v) outreach and (vi) emergency preparedness.

### 5.2.4 Monitoring

Monitoring is carried out to examine the outcome of the activities implementation by comparing the results obtained with the expected results in the planning stage to identify any differences that will be required for management review and continual improvement. Table 6 shows the management practices involved in monitoring the simplified and integrated management system for Responsible Care, including (i) assessment and monitoring, (ii) records and (iii) corrective and preventative actions.

### 5.2.5 Management Review

Management review is the learning process that is produced by the whole process and is used to improve goals, change the methodology or develop tailor-made planning with the results obtained from the whole process. Management review will also be used as a guide for the company to conduct continual improvement. Table 7 shows the management practices involved in management review process

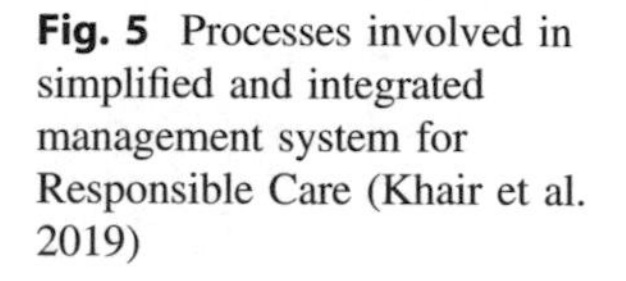

**Fig. 5** Processes involved in simplified and integrated management system for Responsible Care (Khair et al. 2019)

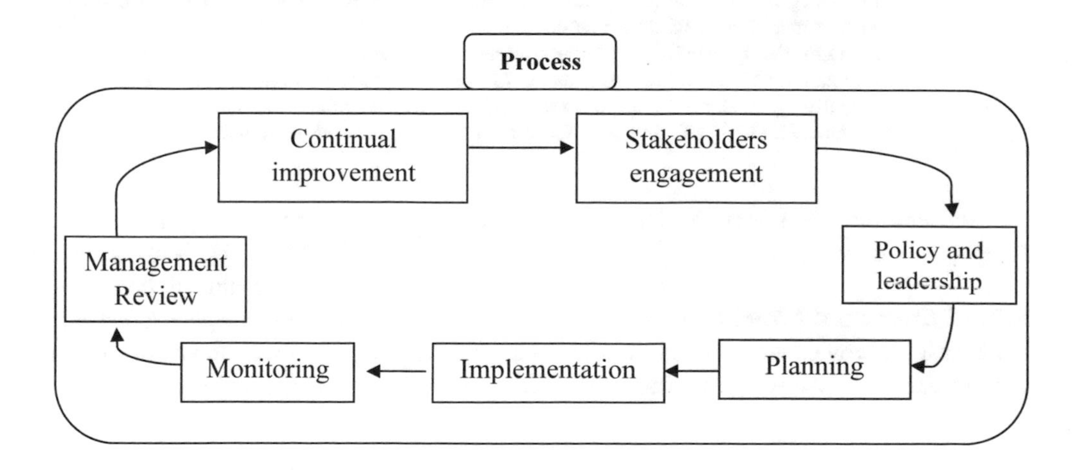

**Table 3** Management practices involved in the policy and leadership process of the simplified and integrated management system for Responsible Care

| Process | Management practices |
|---|---|
| Policy and leadership | To implement the simplified and integrated management system for Responsible Care, the top management of chemical companies shall define the policies used and clearly communicate to all staffs regarding the policies. As such, the top management shall implement and maintain a policy for:<br>• Openness<br>• Environment<br>• Employee health and safety<br>• Process safety<br>• Drugs and alcohol<br>• Product stewardship<br>The company also needs to ensure the commitment of employees to participate in every activity of the company and responsible in implementing the simplified and integrated management system for Responsible Care. Therefore, the top management shall demonstrate their commitment to:<br>• Appointing a representative as the company's ambassador to respond to the concerns of the community and work with the community<br>• Ensuring employees are involved in company's meetings<br>• Ensuring employees are involved in process safety, waste reduction and employee safety programs<br>• Identifying the visions and goals for the implementation of product stewardship<br>• Communicating on company's policies and procedures in distribution activities to the parties handling the products, including distributors |

**Table 4** Management practices involved in the planning process of the simplified and integrated management system for Responsible Care

| Process | Management practices |
|---|---|
| Planning | During the planning stage, the top management shall define objectives and goals, including the importance of fulfilling the codes of management practices. As such, companies shall set goals such as:<br>• Continuous improvement in process safety, targets for waste reduction, goals of implementing product stewardship throughout the company<br>• Developing a crisis management plan<br>• Developing emergency response plan (ERP) within the premise and the vicinity community<br>• Planning the site design, construction and maintenance involved in the process<br>• Planning for complete documentation for hazardous materials<br>The company shall establish, implement and maintain procedures for continuous hazard identification, risk assessment and determination of necessary controls. Therefore, procedures for hazard identification and risk assessment shall take into account:<br>• Identifying the scope of work in every high-risk processes and plant areas<br>• Planning in exposure assessment, process equipment safety analysis, potentially hazardous chemicals and working conditions<br>• Communicating with employees, suppliers, distributors and products receivers about the product hazard risk<br>• Providing EHS information to direct product receivers according to product risk<br>• Designing new or modifying existing products depending on product risk<br>• Developing a list of product hazards<br>• Developing SDS for product risk and distribution<br>• Talking to emergency response authorities about product hazards<br>In addition, company shall consider planning in other areas such as:<br>• Proposing new or modified design of facilities for employee safety and reducing pollution<br>• Taking measures to minimize waste<br>• Setting up early detection of waste releases and discharges<br>• Maintaining a system that integrates EHS impact in product design and process<br>• Preparing procedures for response to accident during distribution<br>• Communicating health and safety information including work tasks with employees |

of the simplified and integrated management system for Responsible Care.

### 5.2.6 Continual Improvement

Continual improvement aims to improve the company's performance by removing unrelated activities, improving efficiency, reducing costs and improving processing capabilities. Table 8 shows the management practices involved in the continual improvement of the simplified and integrated management system for Responsible Care. Continual improvement undertaken shall be in line with the company's policies and goals.

**Table 5** Management practices involved in the implementation process of the simplified and integrated management system for Responsible Care

| Process | Sub-process | Management practices |
|---|---|---|
| Implementation | Documentation | The company shall carry out documentation for:<br>• Safety Data Sheet (SDS) for hazardous processes<br>• Maintenance and inspection of facilities<br>• Change of operation<br>• Employee training<br>• Exposure assessment<br>• Procedure in shipping container selection<br>• Procedures for transporting chemicals in distribution activities<br>• Procedures for unloading chemicals in distribution activities |
| | Selection | The company shall ensure that all parties such as contractors, manufacturers, suppliers, distributors and containers involved in the process are competent, appropriate and experienced. As such, company shall implement and maintain procedures in the selection of:<br>• Contractors who have performance in safety standards<br>• Contractors and manufacturers taking into account waste management practices<br>• Suppliers with procurement policies include guidelines and SDS on their products<br>• Distributors who have the qualifications and emphasize safety and compliance with regulations<br>• Suitable containers used in chemical shipment |
| | Training | The company shall identify the required training that is relevant to the implementation of this management system<br>The company shall provide training or other actions that can meet the requirements of the management system, assess the effectiveness of training provided and maintain relevant records. Therefore, the required training procedures include:<br>• Emergency response especially in community awareness and emergency situations<br>• Communication with the public, the media and local authorities on safety, health and environmental issues<br>• Work safety practices<br>• Proper product handling, operation and emergency response plans<br>• Vehicle drivers to ensure safety in distribution activities |
| | Communication | The company shall ensure that the proper communication process is carried out by the company. As such, companies shall implement and maintain procedures for:<br>• Communication with local authorities about the emergency plan when an incident occurs<br>• Communication with the community about company operations, inventories, risks and response to their concerns<br>• Internal communications with multiple levels and functions within the company by having informal meetings and discussions to respond to their questions and concerns |
| | Outreach | In keeping with the goal of Responsible Care, companies shall establish, implement and maintain procedures for carrying out outreach programs to educate communities, authorities, media and other companies about emergency response programs |
| | Emergency preparedness | The company shall establish, implement and maintain procedures in emergency preparedness and response including:<br>• Risk reduction at all operating stages<br>• Risk reduction and preventive measures in distribution activities<br>• Provision for emergency medical treatment including first aid by qualified providers and occupational medical care for employees at the site<br>The company shall test their procedures periodically to respond in an emergency situation, involving related parties such as:<br>• Providing EHS information to distributors about product risk and handling<br>• Providing EHS information to product receivers matching with product risk, product handling, proper use, recycling, disposal and dissemination of relevant information to downstream users<br>• Safety and system procedures to control incoming and outgoing employees, contractors, visitors, equipment and materials |

### 5.2.7 Stakeholders Engagement

Stakeholders are those directly involved in the company's operations or indirectly in the management process. Table 9 shows the management practices involved in stakeholder engagement of the simplified and integrated management system for Responsible Care. The stakeholders involved in the management process are employees, suppliers, contractors, distributors and product receivers. They shall work together to ensure that the simplified and integrated management system for Responsible Care is implemented

**Table 6** Management practices involved in the monitoring process of the simplified and integrated management system for Responsible Care

| Process | Sub-process | Management practices |
|---|---|---|
| Monitoring | Assessment and monitoring | The company shall establish, implement and maintain procedures to monitor and measure Responsible Care performance on a regular basis. The procedures include:<br>• Risks to employees and local communities<br>• Performance progress in pollution prevention and waste management<br>• Risk in product stewardship<br>• Exposure to employees including hazardous chemicals, physical or biological agents, noise and heat<br>• Employees' health<br>• Health surveillance program<br>The company shall evaluate compliance with other requirements to which they are involved. The procedures include:<br>• Self-assessment in potential impacts from environmental, safety and health in pollution prevention<br>• Community participation in the planning of emergency measures<br>• Vehicle checking and inspection<br>• Qualification of distributors that are compliant with regulations<br>The company shall establish, implement and maintain procedures for recording, investigating and analyzing incidents and follow-up actions in:<br>• Process safety<br>• Distribution<br>• Health and safety of employees<br>Any identified needs for corrective action or any opportunities for preventive action shall be dealt in accordance with corrective and preventative actions.<br>Records on incident investigations shall be documented and maintained |
| | Record | The company shall establish and maintain the required records to demonstrate compliance with the requirements of the simplified and integrated management system for Responsible Care. As such, the company shall have a record:<br>• Latest inventory of hazardous chemicals and waste management<br>• Latest records on the maintenance and inspection recording of facilities<br>• SDS in product stewardship, distribution activity<br>• Employee health<br>• Security program<br>• Exposure assessments |
| | Corrective and preventive action | The company shall establish, implement and maintain procedures to address non-compliance and potential incidents of failure by taking corrective action and preventive measures<br>The company shall ensure that any necessary changes arising from corrective actions and precautions are made in accordance with the simplified and integrated management system for Responsible Care documentation |

effectively. Stakeholders such as local communities, other nearby companies and local authorities are involved in the process of continual improvement of the simplified and integrated management system for Responsible Care whereby the purpose of establishing Responsible Care program itself is to foster a good relationship between chemical companies and the surrounding communities. Hence, feedback from various stakeholders is used to improve the performance of the company and help the company to understand their needs.

## 6 Conclusions

Chemical industries are of major contributors to economic development in Malaysia. However, without good management in chemical industries will pose risks that will not only threaten the quality of the environment, but also human health and safety. With regard to community concerns, chemical industries have introduced Responsible Care program aiming to inculcate chemical-based operators and service providers globally to be more responsible and sensitive to environmental, health and safety issues.

Chemical industries in Malaysia have been implementing Responsible Care program since 1994, promoting environmental, health and safety responsibilities as well as practicing ethical sustainability in chemicals management system. However, the participation of chemical companies in Responsible Care program is still lacking in which only 132 companies are involved although Responsible Care program has been implemented for more than 20 years in Malaysia. Majority of the signatory companies of Responsible Care program are multinational corporations. Though majority of signatory companies agreed that Responsible Care program has improved their company's performance in environmental, health and safety. However, the signatory companies face various challenges in implementing the program, especially in obtaining commitment from all levels of staff.

**Table 7** Management practices involved in the management review process of the simplified and integrated management system for Responsible Care

| Process | Sub-process | Management practices |
|---|---|---|
| Management review | | The company's top management shall review the company's Responsible Care programs implemented within the planned time frame to ensure its suitability, adequacy and effectiveness. Management review shall include assessing opportunities for improvement and the need for changes in management system, including Responsible Care policies and objectives. The records of management reviews shall be retained. Management review shall include information on:<br>• Review in process hazards<br>• Changes in the operation of process<br>• Procedures and practices for process safety<br>• Facilities design in the plant<br>• The responsibilities and job scopes of the employees<br>• Waste management practices and waste minimization practices<br>• Characterization of risks for new and existing products |
| | Frequency | The company shall ensure that the frequency of management review for the simplified and integrated management system for Responsible Care is conducted periodically<br>The company also needs to keep records of periodic assessment results |
| | Communication effectiveness | The company shall review the effectiveness of communications carried out in the process including:<br>• Communication between employees and top management of the company<br>• Communication with the community on emergency response plan, waste and release progress<br>• Communication with local authorities about distribution and product handling activities<br>• Communication with suppliers, distributors and product receivers |
| | Feedback | As one of the measurements of the performance of Responsible Care program, the company shall monitor information related to feedback or suggestions as to whether the company has met the distribution requirements. The methods to obtain and use this information shall be determined |

**Table 8** Management practices involved in the continual improvement process of the simplified and integrated management system for Responsible Care

| Process | Management practices |
|---|---|
| Continual improvement | The company shall continuously improve the effectiveness of the simplified and integrated management system of Responsible Care through the use of policy, objectives, monitoring results, corrective actions and preventive measures and management reviews. The company shall demonstrate procedures to achieve continual improvement including:<br>• Learning process such as sharing information and experiences related to emergencies with other companies in the nearby communities<br>• Identifying the gaps in the management system<br>• Setting the company benchmark for continual improvement<br>• Determining the direction of the company |

**Table 9** Management practices involved in the stakeholder's engagement process of the simplified and integrated management system for Responsible Care

| Process | Management practices |
|---|---|
| Stakeholder's engagement | The company shall involve their stakeholders as their involvement is one of the performance measurements for the simplified and integrated management system for Responsible Care. The company shall identify and confirm their stakeholders, monitor the engagement of its stakeholders whether the company has met the needs of stakeholders<br>The company shall establish, implement and maintain procedures for:<br>• Disclosing company performance in health, safety and environment openly to its stakeholders<br>• Engaging stakeholders including community groups, employees and governments frequently on sustainability, disclosure and performance strategies<br>• Nearby communities to join the company's policy-setting process<br>• Stakeholders such as employees, contractors, suppliers and distributors to be involved in every stage of the process |

Hence, it is proposed to integrate Responsible Care program with other management systems, namely ISO9001, ISO14001 and OHSAS18001 as one of the ways forward to improve Responsible Care program considering averagely signatory companies implement more than one management system in their company. Besides, codes of management practices are also proposed to be simplified for easing the implementation process. In light of the challenges faced by the signatory companies, a simplified and integrated management system for Responsible Care has been developed by incorporating ISO9001, ISO14001 and OHSAS18001 with Responsible Care program. The proposed simplified and integrated management system for Responsible Care is expected to assist chemical companies to improve the implementation of Responsible Care program and encourage more participation of chemical companies in Responsible Care program for the sustainability of chemical industries.

## References

American Chemical Council. (1995). *Responsible Care Annual Report*. United State: American Chemical Council.

American Chemical Council. (2015). Responsible Care Guiding Principles, viewed 22 April 2015 http://responsiblecare.americanchemistry.com/Responsible-Care-Program-Elements/Guiding-Principles/default.aspx.

Asif, M., Searcy, C., Zutshi, A., & Fisscher, O. A. (2013). An integrated management systems approach to corporate social responsibility. *Journal of Cleaner Production, 56,* 7–17.

Athukorala, P., & Menon, J. (1996). Foreign investment and industrialization in Malaysia: Exports, employment and spillovers. *Asian Economic Journal, 10*(1), 29–44.

Belanger, J., Topalovic, M., & West, J. (2009a). *Responsible care: A case study*. Berlin: Walter de Gruyter.

Belanger, J., Topalovic, P., Krantzberg, G., & West, J. (2009b). Responsible care: History & development.

Calcott, P. (2010). Mandated self-regulation: The danger of cosmetic compliance. *Journal of Regulatory Economics, 38*(2), 167–179.

Castleman, B. I. (1997). Responsible care and the Third World. *Environmental Health Perspectives, 105*(1), 16–17.

CEFIC. 2016. The European Chemical Industry Council, viewed 2 December 2016. http://www.cefic.org/About-us/About-Cefic/.

Central Bank of Malaysia. (2013). *Annual Report 2013*. Kuala Lumpur: Central Bank of Malaysia.

Chemical Industry Association of Canada. 2015. Responsible Care Codes, viewed 23 April 2015, http://www.canadianchemistry.ca/responsible_care/index.php/en/responsi/ble-care-codes.

CICM. (1999). *Responsible care code of management practices*. Kuala Lumpur: Chemical Industries Council of Malaysia.

CICM. (2015). Responsible care signatories. http://www.cicm.org.my/2014-04-03-17-08-24/2014-04-03-17-14-17. Accessed on 7 May

CICM. (2019). Responsible care signatories. Chemical Industries Council of Malaysia, viewed 25 January 2019, https://www.cicm.org.my/index.php/responsiblecare/cicmrcprogramme.

Delmas, M., & Montiel, I. (2008). The diffusion of voluntary international management standards: Responsible Care, ISO 9000, and ISO 14001 in the chemical industry. *Policy Studies Journal, 36* (1), 65–93.

Department of Statistics Malaysia. (2011). *Economic Census 2011*. Malaysia: Department of Statistics Malaysia.

Dhara, V. R., & Dhara, R. (2002). The Union Carbide disaster in Bhopal: A review of health effects. *Archives of Environmental Health: An International Journal, 57*(5), 391–404.

Economic Planning Unit. (2013). *Economy History*. Economic Planning Unit, Ministry of Economic Affairs, Malaysia, viewed 18 September 2015, http://www.epu.gov.my.

Evangelinos, K. I., Nikolaou, I. E., & Karagiannis, A. (2010). Implementation of responsible care in the chemical industry: Evidence from Greece. *Journal of Hazardous Materials, 177*(1–3), 822–828.

Federation of Malaysian Manufacturers. (2015). *FMM industry directory chemicals* (2nd ed.). Malaysia: Federation of Malaysian Manufacturers.

Gamper-Rabindran, S. (2006). Did the EPA's voluntary industrial toxics program reduce emissions? A GIS analysis of distributional impacts and by-media analysis of substitution. *Journal of environmental economics and management, 52*(1), 391–410.

Gamper-Rabindran, S., & Finger, S. R. (2013). Does industry self-regulation reduce pollution? Responsible Care in the chemical industry. *Journal of Regulatory Economics, 43*(1), 1–30.

Givel, M. (2007). Motivation of chemical industry social responsibility through responsible care. *Health Policy, 81*(1), 85–92.

Glachant, M. (2007). Non-binding voluntary agreements. *Journal of Environmental Economics and Management, 54*(1), 32–48.

Grolleau, G., Mzoughi, N., & Pekovic, S. (2007). The characteristics of chemical firms registering for ISO 14001 or Responsible Care. *Economic Bulletin, 12*(29), 1–13.

Gunningham, N. (1995). Environment, self-regulation, and the chemical industry: Assessing responsible care. *Law & policy, 17*(1), 57–109.

Hook, G. E. R. (1996). Responsible care and credibility. *Environmental Health Perspect, 104*(11), 1138–1139.

Howard, J., Nash, J., & Ehrenfeld, J. (1999). Industry codes as agents of change: Responsible care adoption by US chemical companies. *Business Strategy and the Environment, 8*(5), 281–295.

JCIA. (2011). *Responsible Care Report 2011*. Japan Chemical Industry Association, Japan, viewed 22 May 2016, http://www.nikkakyo.org/organizations/jrcc/report_e/2011/2011en.pdf.

Jonker, J., & Karapetrovic, S. (2004). Systems thinking for the integration of management systems. *Business Process Management Journal, 10*(6), 608–615.

Karapetrovic, S., & Jonker, J. (2003). Integration of standardized management systems: Searching for a recipe and ingredients. *Total Quality Management & Business Excellence, 14*(4), 451–459.

Khair, N. K. M., Lee, K. E., Mokhtar, M., Goh, C. T., Hanafiah, M. M., Chan, P. W. & Singh, H. (2017). Adoption of Responsible Care program in Malaysian Chemical Industries: Current status and way forward. *Malaysian Journal of Public Health Medicine 2017*, Special Volume (1), 1–6.

Khair, N. K. M., Lee, K. E., Mokhtar, M., & Goh, C. T. (2018). Integrating responsible care into quality, environmental, health and safety management system: A strategy for Malaysian chemical industries. *Journal of Chemical Health and Safety, 25*(5), 10–18.

Khair, N. K. M., Lee, K. E., Mokhtar, M., & Goh, C. T. (2019). Simplified and integrated management system for responsible care (SIMS-RC) in chemical industries. *Management of Environmental Quality: An International Journal, 30*(3), 624–642.

King, A. A., & Lenox, M. J. (2000). Industry self-regulation without sanctions: The chemical industry's responsible care program. *Academy of Management Journal, 43*(4), 698–716.

Lee, K. E., Mokhtar, M., Goh, C. T., Singh, H., & Chan, P. W. (2015). Initiatives and challenges of a chemical industries council in a developing country: The case of Malaysia. *Journal of Cleaner Production, 86,* 417–423.

Lenox, M. J., & Nash, J. (2003). Industry self-regulation and adverse selection: A comparison across four trade association programs. *Business Strategy and the Environment, 12*(6), 343–356.

Lyon, T. P., & Maxwell, J. W. (2011). Greenwash: Corporate environmental disclosure under threat of audit. *Journal of Economics & Management Strategy, 20*(1), 3–41.

MATRADE. (2014). *Chemicals & chemical products.* Malaysia External Trade Development Corporation, viewed 20 May 2015, http://www.matrade.gov.my/en/foriegn-buyers-section/69-industry-write-up-products/519-chemicals-a-chemical-products.

Maxwell, J. W., Lyon, T. P., & Hackett, S. C. (2000). Self-regulation and social welfare: The political economy of corporate environmentalism. *The Journal of Law and Economics, 43*(2), 583–618.

May, C. Y. (2000). *Chemical Industry in Malaysia with Special,* viewed 7 May 2015, http://www.facs-as.org/index.php?page=chemical-industry-in-malaysia-with-special.

MITI. (2010). *Manufacturing*, Ministry of International Trade and Industry, viewed 20 May 2015, http://www.miti.gov.my/cms/content.jsp?id=com.tms.cms.section.Section_44b61991-c0a8156f-628f628f841b8080.

Moffet, J., Bregha, F., & Middelkoop, M. J. (2004). Responsible care: A case study of a voluntary environmental initiative. *Voluntary Codes: Private governance, The Public Interest and Innovation*, 177–208.

Morgenstern, R. D., & Pizer, W. A. (Eds.), 2007. *Reality check: The nature and performance of voluntary environmental programs in the United States, Europe, and Japan*. Resources for the Future.

NZCIC. (2016). *Responsible care New Zealand.* New Zealand Chemical Industry Council, viewed 5 May 2016, http://www.responsiblecarenz.com/Article.aspx?ID=873.

Oskarsson, K., & Von Malmborg, F. (2005). Integrated management systems as a corporate response to sustainable development. *Corporate Social Responsibility and Environmental Management, 12*(3), 121–128.

PACIA. (2015). *Responsible care.* Plastics and Chemical Industries Association, viewed 29 April 2015, http://www.pacia.org.au/programs/responsiblecare.

Phillips, D. C. (2006). *RC14001 certification: An integrated management systems approach.* Quality Digest, viewed 20 May 2015, http://www.qualitydigest.com/may06/articles/04_article.shtml.

Prakash, A. (1999). A new-institutionalist perspective on ISO 14000 and Responsible Care. *Business Strategy and the Environment, 8*(6), 322–335.

Prakash, A. (2000). Responsible care: An assessment. *Business and Society, 39*(2), 183–209.

Rampal, K. G., & Nizam, J. M. (2006). Developing regulations for occupational exposures to health hazards in Malaysia. *Regulatory Toxicology and Pharmacology, 46*(2), 131–135.

RCI. (2015). *Codes of management practices.* Responsible Care Indonesia, viewed 22 April 2015, http://www.responsiblecare-indonesia.or.id/index.php?page=home.

Sabatier, P. A. (1986). Top-down and bottom-up approaches to implementation research: A critical analysis and suggested synthesis. *Journal of public policy, 6*(1), 21–48.

SCIC. 2015. *Responsible care 10 guiding principles.* Singapore Chemical Industry Council, viewed 22 April 2015, http://www.scic.sg/index.php/about-responsible-care.

Shiozaki, Y. (2009). ICCA Responsible Care Leadership Group and Asia Pacific Responsible Care Organization Recent Activities. *The 11th Asia Pacific Responsible Care Conference (APRCC) 2009, International Council of Chemical Association (ICCA)*, Tokyo.

Simon, A., Bernardo, M., Karapetrovic, S., & Casadesus, M. (2013). Implementing integrated management systems in chemical firms. *Total Quality Management & Business Excellence, 24*(3–4), 294–309.

Tapper, R. (1997). Voluntary agreements for environmental performance improvement: Perspectives on the chemical industry's responsible care programme. *Business Strategy and the Environment, 6*(5), 287–292.

Tschopp, D. J. (2005). Corporate social responsibility: A comparison between the United States and the European Union. *Corporate Social Responsibility and Environmental Management, 12*(1), 55–59.

Yohonn, L. (2004). *Moving from responsible care codes to RC 14001 management systems: An analysis of effects on activities that impact regulatory compliance and EHS performance*. M.A. Dissertation: Rochester Institute of Technology.

Yukl, G. (1989). Managerial leadership: A review of theory and research. *Journal of Management, 15*(2), 251–289.

Zeng, S. X., Lou, G. X., & Tam, V. W. (2006). Integration of management systems: The views of contractors. *Architectural Science Review, 49* (3), 229–235.

# Correlating Corporate Social Responsibilities of Chemical Industries in Malaysia Toward Sustainable Development

Noor Syazwani Hassan, Khai Ern Lee, Mazlin Mokhtar, and Choo Ta Goh

**Abstract**

Rapid industrial development causes industries and local communities to be inseparable and interdependent. In line with Malaysia's aspiration to transform to be a developed and high-income nation, manufacturing industry particularly chemical industries have been one of the major contributors to the economic development. Corporate social responsibility is viewed as a model to drive sustainable development of chemical industries in Malaysia to maintain its competitiveness for economic growth, social wellbeing and environmental sustainability. There were only 208 out of 626 chemical companies in Malaysia that have disclosed their corporate social responsibility initiatives in their annual reporting. In this chapter, a list of corporate social responsibility indicators has been gathered and categorized into 12 identified corporate social responsibility components whereby these components can be adopted by chemical companies in Malaysia to further improve the performance of corporate social responsibility by having a clearer guide to protecting the environment and ensuring social progress as well as maintaining economic growth.

**Keywords**

Corporate social responsibility • Sustainable development • Chemical industries • Economic • Social • Environment

## 1 Chemical Industries

Malaysia aims to be a developed and high-income nation by 2020. Hence, the manufacturing industry has been developed into one of the major economic drivers in Malaysia since the 1980s. Malaysia's development have been intensified through various growth programs in the manufacturing industry, such as the Economic Transformation Program launched in 2010. The manufacturing industry is currently the second-largest contributor to the Malaysian economy, accounting for 24.5% in which RM787.6 billion Gross Domestic Product (GDP) was contributed by the manufacturing industry in 2013.

Among the manufacturing industry sectors, chemical industry is one of the most important industries in Malaysia where 626 chemical companies have been recorded by the Federation of Malaysian Manufacturers (FMM), the Chemical Industries Council of Malaysia (CICM) and the Companies Commission of Malaysia (SSM) in Malaysia. According to Lee et al. (2015), there are eight categories of chemical companies in Malaysia, namely oleochemicals, industrial gases, petrochemicals, paints, coating resins, fertilizers, biodiesel and pesticides. Chemical industries have a large network and chemicals produced are used in various sectors in Malaysia, among them are medical sector using chemical drugs; property sector using paints and chemical coating fluids; agricultural sector using chemical fertilizers and pesticides; and other sectors using chemicals in their production activities. As such, it is important for chemical industries to ensure the safety and wellbeing of its stakeholders as an initiative toward sustainable development in Malaysia.

Rapid industrial development causes chemical industries and local communities to be inseparable and interdependent. Various acts and regulations have been gazetted and enforced to control the activities of chemical companies and assist chemical companies in managing operations and chemicals throughout the life cycle, among them are the Environmental Quality Act 1974, Occupational Safety and Health Act 1994, Poisons Act 1952, Factories and Machinery Act 1967 and so forth (Mohamad 2005) to ensure the safety and wellbeing of the environment and human beings. However, corporate social responsibility is required to ensure a better quality of life which protects the

N. S. Hassan · K. E. Lee (✉) · M. Mokhtar · C. T. Goh
Institute for Environment and Development (LESTARI), Universiti Kebangsaan Malaysia, 43600 Bangi, Selangor, Malaysia
e-mail: khaiernlee@ukm.edu.my





K. E. Lee (ed.), *Concepts and Approaches for Sustainability Management*, Advances in Science, Technology & Innovation, https://doi.org/10.1007/978-3-030-34568-6_3

environment and manages natural resources prudently and enhances social progress by engaging stakeholders including communities, customers, suppliers and employees, in addition to prioritizing profitability and compliance with the law in the chemical companies (Gonzalez and Martinez 2004; Matten and Moon 2004; Hilson 2012). Hence, corporate social responsibility is considered as one of the most important concepts to ensure sustainable development of chemical industries (Castka et al. 2004; Mattera et al. 2012; Tayşir and Pazarcik 2013).

## 2 Corporate Social Responsibility

In order to ensure the safety and wellbeing of chemical companies' stakeholders, corporate social responsibility should be taken into account for companies' management (Mohd Radzi et al. 2018a, b, c, d). Friedman (1970) has defined corporate social responsibility in the context of business to increase profit. Bowen and Johnson (1953) claimed that corporate social responsibility should not just to increase profit but should also give a positive impact to the society. Corporate social responsibility is a business obligation to implement policies and make appropriate actions with the goals and values of the society (Bowen and Johnson 1953; Arrow 1973). Many researchers take Bowen and Johnson's definition as a basis in determining corporate social responsibility whereby Jones (1980) defined it as voluntary social responsibility which is the idea that a company has a responsibility to a particular group in the society as well as stakeholders from those recommended by law or by a public contract. Meanwhile, the World Business Council for Sustainable Development is concerned with profits, social and stakeholders where it defined corporate social responsibility as a continuing responsibility of a company to conduct business in an ethical manner and focus on economic growth while improving the quality of life of employees and stakeholders in a company (WBCSD 1999). From the social perspective, Pinney (2001) defined corporate social responsibility as one of the solutions for a company to mitigate the adverse effects of its business activities on the local community as well as to give a positive impact. McWilliams and Siegel (2001) described as social obligations based on voluntary social activity where corporate social responsibility is an act that can strengthen the social positive impact beyond what has been enforced by law. Corporate social responsibility can be classified in terms of stakeholders, social and the environment through the definition by Foran (2001) that corporate social responsibility is the management practices employed by the company for employees and the environment that houses all business activities. From the socio-environmental perspective, Amini and Bienstock (2014) described corporate social responsibility as a company's commitment to continuously improve the quality of society and the environment based on voluntary practice through company activities.

Corporate social responsibility is important for a company whereby the relationship between the company and society is taken into account in the roles, rights and responsibilities of businesses in society (Asif et al. 2013). Corporate social responsibility can be known as a socially responsible company management which can be divided into internal and external corporate social responsibilities (McMurray et al. 2014). Internal corporate social responsibility is related to employee issues and investments in human capital, health and changes to the management. In addition, internal corporate social responsibility is also related to environmental management issues and natural resources, use and production. For external social responsibility, it involves local communities, various stakeholders such as suppliers, customers, local authorities, business partners and non-governmental organizations representing the local community and the environment (Székely and Knirsch 2005). Corporate social responsibility also involves social issues, including benefits offered in terms of safety, health and environmental training, welfare, education schemes, medical benefits and others. Environmental issues also emphasize the preservation and conservation of natural resources such as carrying out recycling activities, noise reduction plans, water treatment processes and compliance with regulatory authorities. Corporate social responsibility strategies in development and programs on social and environmental issues enable companies to establish a close relationship with the community (Sumiani et al. 2007).

With the rapid development of industries, the Prime Minister of Malaysia has announced in the 2007 budget speech that public-listed companies need to disclose their corporate social responsibility activities. Accordingly, the listing requirements were later amended by *Bursa Malaysia* as a new obligation for listed companies to report on the performance of corporate social responsibility in their annual reports. A study conducted by Association of Chartered Certified Accountants (ACCA) Malaysia in 2004, only 43% of companies in Malaysia reported their corporate social responsibility performance and 26% were delayed in their implementation, almost all corporate social responsibility was only manifested in the form of charity (Siwar and Harizan 2009). Given that the manufacturing industry is one of the major contributors to Malaysia's Gross Domestic Product (GDP), industry management should be more accountable in the activities carried out not only because of the demand of the customers and shareholders, but to further enhance their performance in the economic, social and natural aspects around (Givel 2007; Hassan et al. 2018a).

## 2.1 Levels of Corporate Social Responsibility

To perform corporate social responsibility, a company needs to make profits, comply with the law and help the society in the form of funds, time and skills. According to Carroll (1983), corporate social responsibility is undertaken in a business entity to benefit economically, compliance with the law and social support. A pyramid that encompasses four levels of corporate social responsibility has been proposed (Carroll 1979; Geva 2008) based on economic, legal, ethical and philanthropic responsibilities. These four levels clearly set the goal for each level of corporate social responsibility in Fig. 1.

- Economic responsibility: The company gains profit in return of offering products to the community, meeting the needs of stakeholders and promoting innovation.
- Legal responsibility: The company needs to do business in compliance with the law.
- Ethical responsibility: The company needs to have moral, act fair, respect for the people's rights, avoid danger or harm caused by others.
- Philanthropic responsibility: The company needs to carry out any beneficial activities for the society in the form of individuals or organizations.

## 2.2 Conflicts of Corporate Social Responsibility

The four levels of corporate social responsibility mentioned earlier are interconnected among each other whereby the balance between economic, legal, ethical and philanthropic responsibilities toward various stakeholders is needed to show different values and beliefs between a company and another (Moon 2007). However, there are also conflicts in corporate social responsibility in which they can be divided into economic and non-economic conflicts (Lantos 2001). For economic conflicts, stakeholders involved are shareholders and users. Conflicts involving consumers are rising prices to cover the rising costs, safe product manufacturing and user information. At the same time, conflicts occur when there is a demand to improve financial performance by the investors whereby the management of the company has a responsibility to their investors to obtain maximum return on investment (Lantos 2001).

Non-economic conflicts include legal, ethical and philanthropic aspects often arise from stakeholders, such as workers, local communities and the environment. Among, the conflicts are their demands for companies to focus resources on corporate social responsibility. For workers, they often complain about workers' rights at work, including provisions for health and safety of workers, discrimination in employment opportunities, performance-based salary, dismissal and promotion, zero career termination, stock ownership by employees and vacation programs for families (Lantos 2001). Conflicts in the community and the environment are greater than the conflicts of stakeholders and workers (Lantos 2001) where they demand that business operations to be carried without threatening the safety of the society, financial assistance should be given to minority housing areas, special training and necessary employment opportunities to be provided for unemployed, investment in equipment that can reduce pollution, contribute to welfare and non-profit organizations and provide non-compensation services for executives or other non-business assignments.

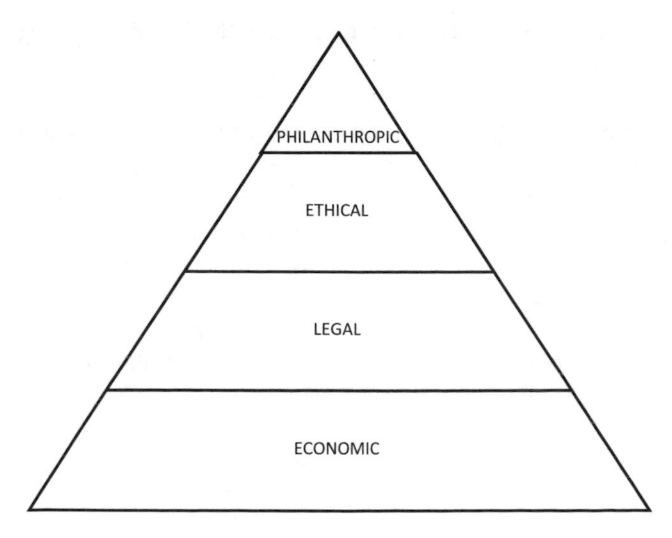


**Fig. 1** Levels of corporate social responsibilities (Carroll 1999)

## 2.3 Benefits of Corporate Social Responsibility

A company will receive positive impacts if they perform corporate social responsibility in their business. Internal and external benefits may exist for companies performing corporate social responsibility in which it can be divided into four types of benefits, i.e. economic, legal, ethical and philanthropic benefits.

### 2.3.1 Economic Benefits

(i) Creating Financial Value

According to Naser et al. (2002), he found that there was a positive relationship between financial and internal performances with corporate social responsibility. Some steps need to be taken to perform corporate social responsibility, i.e., collecting data, such as collecting and analyzing data on resources and materials usage and evaluating business processes. This process will help companies to identify opportunities for cost savings and generate income through the efficient use of resources and materials. For example, DuPont has reduced the cost of $ 2 billion from energy used since 1990 (Porter and Kramer 2006). This is the economic outcome of a wise business corporate social responsibility decision other than benefiting from the environment.

(ii) Increasing the Competitive Advantage

There are many companies using corporate social responsibility as a strategy to differentiate their brands, products or services from other companies. Consumers are more likely to buy a product and use the service of a company that operates without neglecting corporate social responsibility. In addition, customers have the option to deposit their money into a socially responsible company. Hence, this is important because corporate social responsibility in the company will help increase the level of consumer confidence in their products and services. According to Ameer and Othman (2012), companies that emphasize corporate social responsibility and sustainability have better financial performance and competitiveness than those who have no such initiative.

(iii) Encouraging Innovation

Corporate social responsibility also helps companies in innovation. In the process of developing corporate social responsibility, there are many opportunities that can stimulate new ideas and creativity, and this helps to enhance the company's innovation. For example, Toyota's concern over the impact of their car production on the environment has resulted in innovative hybrid car models—Toyota Prius (electric hybrid/petrol) that are innovative and can compete with other car manufacturers and benefit the environment (Porter and Kramer 2006).

(iv) Attracting Potential Investors

By executing corporate social responsibility, it can convince potential investors to invest in the company. In managing company, it includes investor's engagement including consideration of corporate social responsibility. The management provides a mechanism to ensure that the company reports details to the stakeholders about the information required for assessment and evaluation (Weber 2008).

(v) Brand Differences

Corporate social responsibility can be used by companies to differentiate themselves from competitors in sales and marketing purposes. For example, a successful company using corporate social responsibility as a brand differentiator is Ben & Jerry and The Body Shop (McWilliams and Siegel 2001) where their concept of corporate social responsibility has led to their brand-winning competition.

### 2.3.2 Legal Benefits

(i) Reducing Corporate Risk

Accidents and any problems that occur in the company may give negative impact on the reputation of the company. Corporate risk can be reduced by implementing appropriate rules and regulations through corporate social responsibility in a company (Weber 2008). The process of decision-making and management of corporate social responsibility help a company to gain a better understanding of the risks and take appropriate action. An effective corporate social responsibility program can also help in avoiding offenses and adverse events.

(ii) Improving the Reputation of the Company

Companies that practice corporate social responsibility as the core elements of management and take into account of stakeholders in the management of corporate social responsibility can enhance the company's image (Brønn and Vrioni 2001). According to Porter and Kramer (2006), Nestlé's approach to working with small farmers is to enhance the reputation of the company. Not only that, but it also has a positive impact on developing countries by the company. In addition, the implementation of corporate social responsibility can help companies to manage property trustees, perceptions and protect the reputation of the company as stakeholders' perceptions are important in influencing the company's economic performance.

### 2.3.3 Ethical Benefits

(i) Motivating, Coordinating Existing Staffs and Attracting New Talents

A strong corporate social responsibility policy can be used to attract and retain employees. It can also help to maintain pride of the organization (Lougee and Wallace 2008). Employees will be more confident in working under an ethical company focusing on safety and environmental responsibilities. This will increase employees' loyalty and reduce the number of employee accidents, especially at high risk and hazardous chemical companies. According to Witt (1991), employees will give them greater commitment when they are treated fairly, and employees tend to identify fair procedures because they will benefit from the procedure and they will be compared with other companies as an example that treats workers fairly.

(ii) Demonstrating Transparency

Companies that perform corporate social responsibility are more transparent than any other companies that do not as they manage environmental, social and economic risks in ways that enable consumers to trust in their products. Some companies show their transparency by disclosing their corporate social responsibility activities to the public and demonstrating how they manage their companies toward sustainable development. For example, 19 major oil and gas and mining companies agree to reject bribes through full public disclosure and report all corporate payments to the governments in the countries in which they operate (Porter and Kramer 2006).

### 2.3.4 Philanthropic Benefits

The public's perception has the greatest effect when operating companies in highly regulated and extractive industries, e.g., mining and chemical manufacturing (Porter and Kramer 2006). Communities tend to be positive about a company if the company does not neglect the interests and welfare of the society as a result of their business activities. Thus, by establishing relationships between companies and communities, the community's perspective on the activities of the company will improve.

## 3 Corporate Social Responsibilities Toward Sustainable Development

Corporate social responsibility and sustainable development are still disputed concepts and are often debatable (Moon et al. 2005). Corporate social responsibility and sustainable development have their own values and the concepts are not considered solely empirical. Therefore, the principles of using corporate social responsibility and sustainable development are quite open (Moon 2007). According to Hart (1995), there are two major sustainable development strategies that are pollution prevention and environmental impact minimization. In order to achieve these strategies, a company should adopt corporate social responsibility in their activities aiming at reducing pollution by using continuous methods and focusing on a clear goal rather than a cost-effective method of preventing pollution (Rooney 1993).

### 3.1 Concept

In this chapter, corporate social responsibility is deliberated based on the concept proposed by Carroll (1979, 1991), while sustainable development is deliberated based on the three dimensions of social, environment and economy. The component of corporate social responsibility toward sustainable development can be described in combinations that relate the levels of corporate social responsibility to sustainable development (Hassan et al. 2018b). Figure 2 shows a concept that maps corporate social responsibility toward sustainable development where there are 12 identified corporate social responsibility components, namely economy, environment-economy, social-economy, economy-legal, environment-legal, social-legal; economic-ethic, environmental-ethic, social-ethic, economy-philanthropy, environment-philanthropy and social-philanthropy.

#### 3.1.1 Economic Responsibility Toward Sustainable Development

Among the four levels of corporate social responsibility, economic responsibility is the most important. Carroll (1991) and Poitras (1994) agreed that a company is considered consistent when they conduct business according to social behavior that can increase stock earnings, maximize productivity, enhance competitiveness and have high level of effectiveness in business. For large companies with excessive resources, it enables the companies to engage in various innovative projects that can enhance the companies' competitiveness and economy (Waworuntu et al. 2014).

| Levels of CSR | | Social | Environment | Economy |
|---|---|---|---|---|
| Philanthropy | → | Social-Philanthropy | Environment-Philanthropy | Economy-Philanthropy |
| Ethic | → | Social-Ethic | Environmental-Ethic | Economy-Ethic |
| Legal | → | Social-Legal | Environment-Legal | Economy-Legal |
| Economy | → | Social-Economy | Environment-Economy | Economy |

**Fig. 2** Concept of correlating corporate social responsibility toward sustainable development

However, for companies with a lack of resources, the companies are more likely to give priority to activities that can improve profit and profit-oriented activities rather than engaging in corporate social responsibility activities (Roberts 1992). Therefore, companies need to identify their interests in society and adopt their economic, legal, ethical and philanthropic responsibilities in their business activities. Otherwise, sustainable development will not be achieved (Lindgreen and Swaen 2005; Luo and Bhattacharya 2006).

Financial performance including income and assets should be reported as an economic responsibility to shareholders as stressed by Friedman (1970) that the most important role in business activities is to maximize revenue. According to Carroll and Shabana (2010), profits in business have three main objectives based on classic economic outlook: first, to identify the effectiveness of business operations; second, as the "risk premium" required by the company to remain in the business; and finally, to protect future stocks. Economic responsibility is not only to measure profits from business activities, but it also measures the minimum amount that must be produced by the company.

For environmental-economic responsibility, the efficiency of natural resources and recycling activities are emphasized. The amount of water and energy used, renewable energy, water recycling and waste are monitored to maintain high level of environmental efficiency because environmental-economic responsibility is one of the company's efforts to ensure the reduction of environmental risks and to maximize revenue (Pinney 2001).

The World Business Council for Sustainable Development (WBCSD 1999) cares about revenues, communities and stakeholders in which socioeconomic responsibility emphasizes corporate social responsibility as the company's ongoing responsibility to operate ethically and focus on economic growth while improving the quality of life of their employees and shareholders. The involvement of all stakeholders in the process of planning, operation and communication with shareholders as well as corporate information sharing is an important aspect of socioeconomic responsibility, not only to increase profits but also to take into account of social aspects in management (Bowen and Johnson 1953).

### 3.1.2 Legal Responsibility Toward Sustainable Development

Carroll (1991) noted that legal responsibility is to conduct business in a consistent management of compliance with laws and regulations. For companies that generate lucrative profits, it is important for the companies to carry out legal responsibility in their business activities at least to meet the minimum legal requirements. The law can be classified into three categories, namely compliance, avoiding civil lawsuits and expecting the law (Schwartz and Carroll 2003).

According to Schwartz and Carroll (2003), companies need to avoid any current or future civil court claims that may be due to negligence in business activities because they are aware that they may be charged over their negligence. Therefore, companies need to implement strategies by complying with the law in their business operations to avoid any possible legal action. Reporting products, services and labeling are the measures to prevent economic-legal action from happening (Carroll 1979). The community also hopes that business operations can play an effective role, in accordance with the laws and regulations that a business should take to operate (Carroll 1979).

Environmental-legal responsibility can be explained by Tsoutsoura (2004) where greater success can be achieved by respecting the ethical norm, people, society and nature. There are four aspects that are typically reported for environmental-legal responsibility, namely environmental policy and compliance, waste management, water quality and air quality. It is important to undertake climate change risk assessment, life cycle assessment, environmental policy optimization, environmental policy compliance and material abuse policy. In terms of waste management, environmental-legal responsibility covers hazardous and harmless waste management. Water quality includes the amount of treated water released, the amount of freshwater used, the amount of volatile organic carbon released, the amount of phosphorus in the wastewater and the amount of nitrogen in the wastewater, and the air quality includes the amount of $CO_2$ and nitrogen emissions. Positive image can be shown to the public by revealing environmental performance data. All these aspects are based on the importance of different issues in a business depending on the type of company in which each business is concerned with different aspects (Carroll 1979).

Socio-legal responsibility is purely social action, beyond the law enforcement. This can be explained by Bowen (2013) that corporate social responsibility is the responsibility of an organization to implement a set of policies, determine decisions or take appropriate measures in terms of goals and respect for local communities. Safety and health aspects covering the number of fatalities, the number of employees lost during the event, the reduction and prevention of accident and safety, health hazard assessment and risk assessment are the socio-legal responsibility of a company. In addition, socio-legal responsibility also includes compliance with labor law as a company is responsible for their own employees.

### 3.1.3 Ethical Responsibility Toward Sustainable Development

Ethical norm is known as equity and justice, while ethical responsibility refers to any action and behavior that are expected or restricted by community groups even though they are not coded in the law. This is supported by Smith and Quelch (1993) where ethic is moral action, doing the right thing, respecting the right of the people fairly, morally and preventing damage or any social injury and damage caused by others. Carroll (1991) noted that there is a need for a company to operate in a way that is consistent with the norm of the society to recognize and respect ethical norms as well as achieve corporate goals. This is important for a company that is known and respected by the society for conducting moral and ethical activities that are not based solely on laws and regulations. In this case, it is important for a company to carry out moral norms and values in business activities to each individual (Tayşir and Pazarcik 2013). To ensure ethical employee behavior, ethical responsibility forces a company to perform moral duties in management as it is part of the company's business ethics (Velasquez and Velazquez 2002). As such, the companies should provide codes of practice and work at all levels as ethical responsibility to ensure that decisions are made ethically.

Economic-ethical responsibility can be defined as an operation that is considered as an ethical act based on a theory that offers benefits to the community or other actions aimed at improving economic benefits, e.g., reducing the net cost to the community (Velasquez and Velazquez 2002). This economic-ethical responsibility can be attributed to business's "emerging ethics," described by Reidenbach and Robin (1991) as immediate company management to seek a balance between profit and ethic. Therefore, employees' salaries and bonuses are essential to avoid standard ethics from being ignored in the process of achieving the company's goals.

Environmental-ethical responsibility is a reaction and action related to environmental management (Strand 1983). Hence, reports and management plans are important to identify the values of companies and ethics beyond what is listed in the laws and regulations. The common aspects reported under environmental-ethical responsibility are the occurrence and potential hazard, industrial hygiene, crisis management plan, pollution control and ethical-based product execution and surveillance in which the company should produce safe products to the environment (Carroll 1979).

Carroll (1991) noted that socio-ethical responsibility refers to standards, norms or beliefs that carry out responsibility for employees, shareholders and treat the community fairly, equally or appropriately to respect or protect the moral rights of stakeholders. There are two main aspects that have been taken into account, i.e., management policies and workers' rights, to identify and value new or emerging ethical norms that are used by society and operate in accordance with social and ethical norms. Those aspects include no child labor, no forced labor, code of action, no gift policy, HIV/AIDS policies and information disclosure policies. While the rights of workers are the measurement of feedbacks and actions on employees, employee rewards and recognitions, diversity based on gender, age and region, employee benefits, number of employees, employee training, privacy and equal opportunities, health programs for their employees and families, transparency and respect for human rights.

### 3.1.4 Philanthropic Responsibility Toward Sustainable Development

Based on the definition by Carroll (1991), philanthropic responsibility is to conduct business consistently with charitable practices and contribute to the community, help to improve the company's image, and even important for managers and employees of companies to participate in voluntary and charitable activities in the community, provide assistance to voluntary organizations such as private and public education to assist projects that can improve the quality of life of the community. Corporate social responsibility is often a controversy when many assume that voluntary activities can restore time and money. Abdulrazak and Ahmad (2014) found that charity became the most common activities carried out through corporate social responsibility in Malaysia causing people to incline to misunderstand that corporate social responsibility is solely donation without knowing that donation is just part of the philanthropic responsibility in corporate social responsibility.

Lougee and Wallace (2008) defined economic-philanthropic responsibility as obligations committed to local communities, customers and employees who are not negligible, at least acting fairly by giving priority to the company's economic objectives that create profit for the company's shareholders. The type and amount of philanthropic activities reported by companies in Malaysia are one of the ways of economic-philanthropic responsibility that can have a positive impact on the society. Bruch (2005) noted that many companies use philanthropy in the form of financial to enhance corporate image and competitiveness.

Environmental-philanthropic responsibility has been taken into account as a company's commitment to improve the quality of the environment through their ongoing voluntary activities (Amini and Bienstock 2014). The environmental-philanthropic programs undertaken by Malaysian companies are green building initiatives, greenhouse gas reduction programs, environmental conservation efforts, environmental sustainability activities, biodiversity and animal welfare initiatives. These voluntary initiatives can provide environmental awareness to workers and communities to achieve sustainable development.

According to Carroll (1991), socio-philanthropic responsibility is a responsibility that demonstrates the conduct of a company or appropriate action to social wellbeing.

In terms of achievement, it is clear that the company must determine and fulfill social expectations and program objectives by involving the community which is one of the important aspects of socio-philanthropic, namely to improve the quality of life of the community and the relationship between the company and the local community through community involvement such as voluntary community programs.

## 4 Case Study: Corporate Social Responsibility of Chemical Industries in Malaysia

According to Teoh and Thong (1984), there is evidence of corporate social responsibility implementation in Malaysia since the 1980s. However, after more than a decade, corporate social responsibility only shows their progress. Due to the policies and initiatives by the federal government and its agencies (Sharma 2013), there are two main drivers in the progress of corporate social responsibility in Malaysia. First, the "Silver Book" launched in May 2005 is a set of principles and guidelines of corporate social responsibility for government-linked companies (GLCs) in Malaysia (Lu and Castka 2009) whereby its objective is to assist GLCs in applying corporate social responsibility into their business activities (Abdulrazak and Ahmad 2014). Second, Bursa Malaysia requires all public-listed companies (PLCs) in Malaysia to disclose their corporate social responsibility initiatives in the companies' annual financial reports (Yam 2012). Since then, various corporate social responsibility awards have been introduced and one of them is the Prime Minister's Corporate Social Responsibility Award which was launched in 2007 (Abdulrazak and Ahmad 2014). A study was conducted and found that, between July 2003 and December 2004, Malaysian companies have donated more than RM 82 million in charitable activities as their corporate social responsibility (Prathaban and Rahim 2005).

However, most companies only focus on aspects, such as community (Darus et al. 2013), charity, human resources (Yam 2012), products and the environment (Saleh et al. 2010). In fact, corporate social responsibility practices are still unable to meet the expectations of the local communities (Amran et al. 2013). The company's corporate social responsibility should include all levels of corporate social responsibility, i.e., economy, legal, ethic and philanthropy as proposed by Carroll (1979).

Besides that, studies show that there is a positive relationship between corporate social responsibility and company performance (Fuzi et al. 2012; Mustafa et al. 2012; Ahamed et al. 2014). This can also be attributed to the type of company ownership. According to Hoq et al. (2010), corporate social responsibility and company ownership show a positive relationship in which large companies tend to disclose corporate social responsibility activities that can help to improve the companies' financial performance. However, not all the aspects studied cover all levels of corporate social responsibility, i.e., economy, legal, ethic and philanthropy. For PLCs, those aspects have been set by Bursa Malaysia, namely the environment, workplace, community and market. There are also various limitations, such as Bursa Malaysia often championed corporate social responsibility as the key to sustainability. However, there is no specific requirement for companies to report their sustainability activities in their corporate social responsibility report. The corporate social responsibility or sustainability reports reported by the PLC show a rewarding performance but majority of the companies in Malaysia have little knowledge and awareness on sustainability issues involving their business activities.

Although rapid economic development has spurred the growth of chemical companies which also contributed to Malaysia's Gross Domestic Product (GDP) (Lee et al. 2015), chemical companies often receive perception of being selfish and making profit in every activities and not it is puzzled if a chemical company can succeed in investing in the community and the environment. Sustainable development is essential to improve chemical companies' production process, ensure justice and achieve sustainability in improving industry competitiveness (Hove 2004). As such, corporate social responsibility is a response to the negative perceptions of chemical industries.

Van Marrewijk (2003) related corporate social responsibility and sustainable development as a voluntary social and environmental engagement in companies' activities with stakeholders to emphasize the importance of environment, social, volunteerism, stakeholder interests and profits in a company's business. While the Malaysian government has made it compulsory for all PLCs to carry out their corporate social responsibility disclosure to the public as one of the initiatives toward sustainable development, both corporate social responsibility and sustainable development concepts used by the industry are considered contradictory and unclear. Hence, there is a need to map corporate social responsibility toward sustainable development of chemical companies in Malaysia.

## 5 Dimensions, Components and Aspects of Corporate Social Responsibility Toward Sustainable Development

Based on the concept of correlating corporate social responsibility toward sustainable development as shown in Fig. 2, annual reports were collected from 208 chemical companies out of the total number of 626, to identify the corporate social responsibility aspects covered by the chemical companies. A list of aspects of corporate social responsibility has been reported by 208 chemical companies in Malaysia in which those aspects can be categorized into 12 identified corporate social responsibility components, namely economy, environment-economy, social-economy, economy-legal, environment-legal, social-legal, economic-ethic, environmental-ethic, social-ethic, economy-philanthropy, environment-philanthropy and social-philanthropy. The aspects of corporate social responsibility are deliberated according to the dimensions of sustainable development.

### 5.1 Economy

As shown in Table 1, corporate social responsibility aspects in economy dimension are divided into charitable contributions, employee benefits, products and services and financial performance. Charitable contribution is intended to give a positive impact to the society whereby this aspect is reported based on the type and amount of donation. While employee benefits such as salaries and bonuses are reported under the component of economy-ethics, aiming to avoid ethical norms from being compromised in achieving corporate goals. Products and services labeling are the economy-legal responsibility of the company to provide goods and services that meet the legal requirements. Financial performance reporting revenue, property and asset, committing to be as profitable as possible and implementing consistent ways to maximize share earnings and maintain a strong competitive position.

### 5.2 Social

As shown in Table 2, corporate social responsibility aspects identified under social dimension are community engagement, management policy, employee rights, safety and health monitoring and law and stakeholder engagement. Under community engagement, the type of community programs, the number of community programs and the number of volunteers for community programs are reported to showcase how the company improves the quality of life of the community and enhances the relationship between the company and the community. Management policy and employee rights are to recognize and respect new or evolving ethical moral norms adopted by the society and to perform in a manner which is consistent with the society ethical norms. Aspects for management policy are no child labor, no forced labor, policy code of conduct, no gift policy, policy on HIV/AIDS and whistle blowing policy, whereas for employee rights, measurement, feedback and action on employees, employee training, privacy and equal opportunities, diversity of gender, age and race, number of employees, respect human right, wellness program for employees and their families as well as transparency are reported. Aspects for safety and health monitoring and law reported are the number of fatalities, employees' "lost-time" incident, safety and health hazard recognition and risk assessment, mitigation and prevention action and labor law compliance, as a guide to ensure law-abiding practices. For stakeholder engagement, the aspects are reporting shareholder information, corporate information and engage all stakeholders in the company, which are intended to gather all information and ideas from stakeholders in a company.

### 5.3 Environment

Table 3 shows the corporate social responsibility aspects identified under environment dimension, namely environmental programs, reporting and management plans, environmental policy and compliance, waste management, water quality, air quality, resource efficiency and recycling activities. Environmental programs include green building, program to reduce greenhouse gases, environmental conservation initiatives, implementing and maintaining environmental sustainability, biodiversity conservation initiatives and animal welfare aimed at providing environmental awareness to workers and communities. Report and

**Table 1** Corporate social responsibility aspects for economy dimension

| Components | Aspects |
|---|---|
| Economy-philanthropy | Charitable contribution<br>– Type of donation<br>– Amount of donation |
| Economy-ethic | Employee benefits<br>– Employees' salaries and bonuses |
| Economy-legal | Products and services<br>– Labeling |
| Economy | Financial performance<br>– Revenue<br>– Property and assets |

**Table 2** Corporate social responsibility aspects for social dimension

| Components | Aspects |
|---|---|
| Social-philanthropy | Community engagement<br>– Type of community programs<br>– Number of community programs<br>– Number of volunteers for community programs |
| Social-ethic | Management policy and employee rights<br>(a) Management policy<br>– No child labors<br>– No forced labor<br>– Policy code of conduct<br>– No gift policy<br>– Policy on HIV/AIDS<br>– Whistle blowing policy<br>(b) Employee rights<br>– Measurement, feedback and action on employees<br>– Employee training<br>– Privacy and equal opportunities<br>– Diversity (gender, age, race)<br>– Number of employees<br>– Respect human right<br>– Wellness program for employees and their families<br>– Transparency |
| Social-legal | Safety and health<br>(a) Safety and health monitoring<br>– Number of fatalities<br>– Employees' "lost-time" incident<br>– Safety and health hazard recognition and risk assessment<br>– Mitigation and prevention action<br>(b) Law<br>– Labor law |
| Social-economy | Stakeholders engagement<br>– Shareholders info<br>– Corporate info<br>– Engage all stakeholders in the company |

management plan include reporting of incident and potential hazard, industrial hygiene, crisis management plan, implementing pollution control and product stewardship aimed at recognizing corporate and ethical integrity beyond compliance with laws and regulations. Environmental policy and compliance cover climate change risk assessment, life cycle assessment optimizing environmental policy and environmental compliance and substance misuse policy. Waste management includes total hazardous waste and non-hazardous waste production. Water quality includes total freshwater used and treated water discharged, total organic carbon and volatile organic carbon and total phosphorus and nitrogen contained in wastewater while air quality includes total $CO_2$ and nitrogen emission, to give a positive image to the government and showcase a successful company that fulfills legal obligations in the companies' environmental activities. Resource efficiency and recycling activities include total water consumption and energy used as well as total renewable energy used and water and waste recycling, to enhance the companies' operational efficiency.

## 5.4 Stakeholders Engagement

There are two categories of stakeholders, i.e., primary and secondary stakeholders as shown in Table 4. The relationship between the companies and their stakeholders is very important to bring corporate social responsibility toward sustainable development (Isa 2012). As noted by Bowen and Johnson (1953), corporate social responsibility is not only to increase profit but also to give impact to the society while sustainable development is a strategy in which society gains economic development, which also benefits the environment and the quality of life (Mawhinney 2008). By deliberating both definitions, both corporate social responsibility and sustainable development share the same goal of increasing the growth of the economy, protecting the environment and improving the quality of life of the society in which both definitions emphasize the interests of stakeholders.

The primary stakeholders are those who have a direct relationship with a company in realizing the mission to produce a product or service for customers. Secondary

**Table 3** Corporate social responsibility for environment dimension

| Components | Aspects |
|---|---|
| Environment-philanthropy | Environmental programs<br>– Green building<br>– Program to reduce greenhouse gases<br>– Environmental conservation initiatives<br>– Implementing and maintaining environmental sustainability<br>– Biodiversity conservation initiatives<br>– Animal welfare |
| Environment-ethic | Reporting and management plan<br>– Reporting of incident and potential hazard<br>– Industrial hygiene<br>– Crisis management plan<br>– Implementing pollution control<br>– Product stewardship |
| Environment-legal | Environmental policy and compliance<br>(a) Policy and compliance<br>– Climate change risk assessment<br>– Life cycle assessment optimizing environmental policy<br>– Environmental compliance and substance misuse policy<br>(b) Waste management<br>– Total hazardous waste<br>– Total non-hazardous waste<br>(c) Water quality<br>– Total freshwater used and treated water discharged<br>– Total organic carbon and volatile organic carbon<br>– Total phosphorus and nitrogen in wastewater<br>(d) Air quality<br>– Total $CO_2$ emission and nitrogen emission |
| Environment-economy | Resources efficiency and recycle activities<br>(a) Resource efficiency<br>– Total water consumption and energy used<br>– Total renewable energy used<br>(b) Recycling activities<br>– Water and waste recycling |

**Table 4** List of primary and secondary stakeholders

| Stakeholders | |
|---|---|
| Primary | Secondary |
| (i) Employees<br>(ii) Consumers<br>(iii) Retail customers<br>(iv)Suppliers<br>(v) Shareholders<br>(vi) Investors<br>(vii) Debt holders<br>(viii) Stockholders<br>(ix) Contractors | (i) Government<br>(ii) Academia<br>(iii) Non-governmental organizations<br>(iv) Media/journalists<br>(v) Scientific organizations<br>(vi) Industry association<br>(vii) Consortia<br>(viii) Labor association<br>(ix) Non-profit organization<br>(x) Regulator<br>(xi) Standardization bodies<br>(xii) Rating organizations<br>(xiii) Sustainability organizations<br>(xiv) Politicians<br>(xv) Intergovernmental organizations<br>(xvi) Community<br>(xvii) Public |

stakeholders include social and political organizations that support the mission of a company for certain reasons and provide approval for activities (Haniffa and Cooke 2005). These parties are communities, local governments and non-governmental organizations (NGOs) (Park et al. 2014). The development and implementation of corporate social responsibility can be considered as organizational change process for a new step in administering the company in the future. It is to coordinate organizations with businesses based on dynamic demand and more appropriate environment by identifying and managing stakeholders (Gordon et al. 2012). Apart from taking into account environmental issues, relationships between stakeholders are also important in the company, especially cooperation in the event of a crisis in the company. Therefore, corporate social responsibility should take steps to create interaction, positive relationships and persistent communication with stakeholders to build confidence and social learning among them (Zhao et al. 2012). According to Mitchell et al. (1997), stakeholders can be grouped using three main attributes of power, legitimacy and urgency. Generally, the company will pay more attention to its higher stakeholders (Zhao et al. 2012). These features will vary from time to time depending on the structure of the company. Dahl (1957) defined power as a "relationship between social organization in which a social organization, A, can get another social organization, B, to do something B cannot do and vice versa." In addition, Suchman (1995) described legitimacy as a general perception or assumption that the action of a person is appropriate, correct or appropriate in a social system constructed from norms, values, beliefs and definitions. For the third attribute, the Merriam-Webster Dictionary defined urgency as "calling attention immediately emphasized" (Zhao et al. 2012). Based on that definition, "power" may refer to the government, "legitimacy" may refer to shareholders and "urgency" may refer to the society and the environment. All these three features will help to identify stakeholders in the company (Mitchell et al. 1997).

## 6 Conclusions

Malaysia aims to transform to be a developed and high-income nation and the manufacturing industry has been developed into one of the major economic drivers in Malaysia since the 1980s. Rapid industrial development causes industries and local communities to be inseparable and interdependent. Chemical industries particularly have been one of the major contributors to the GDP in Malaysia whereby there are 626 chemical companies registered with the Federation of Malaysian Manufacturers (FMM), the Malaysian Chemical Industry Organization (CICM) and the Companies Commission of Malaysia (SSM). Hence, sustainable development of chemical industries in Malaysia is vital to maintain its competitiveness for economic growth, and corporate social responsibility is considered as one of the most important concepts to ensure sustainable development of chemical industries.

Corporate social responsibility has emerged to ensure a better quality of life which it protects the environment and manages natural resources prudently and enhances social progress by engaging stakeholders in the companies' activities. Malaysian corporate social responsibility has been growing since the launch of "Silver Book" in 2005 that sets the principles and guidelines of corporate social responsibility for government-linked companies (GLCs). However, corporate social responsibility and sustainable development concepts used by the industry are often debatable due to the contradicting interpretations by the industry. Hence, there is a need to map corporate social responsibility toward sustainable development of chemical companies in Malaysia.

There are only 208 out of 626 chemical companies in Malaysia that have disclosed their corporate social responsibility initiatives in the annual report. The correlating exercise sorted a list of aspects of corporate social responsibility that have been reported by 208 chemical companies in Malaysia. Those reported aspects can be categorized into 12 identified corporate social responsibility components, namely economy, environment-economy, social-economy, economy-legal, environment-legal, social-legal; economic-ethic, environmental-ethic, social-ethic, economy-philanthropy, environment-philanthropy and social-philanthropy. The identified aspects of corporate social responsibility can be referred and adopted by chemical companies in Malaysia to further improve the performance of corporate social responsibility regardless of the size of the company. With the correlating of corporate social responsibility toward sustainable development, chemical companies have a clearer guide to protect the environment and ensure social progress which is in line with the economic growth.

## References

Abdulrazak, S. R., & Ahmad, F. S. (2014). The basis for corporate social responsibility in Malaysia. *Global Business & Management Research*, *6*(3).

Ahamed, W. S. W., Almsafir, M. K., & Al-Smadi, A. W. (2014). Does corporate social responsibility lead to improve in firm financial performance? Evidence from Malaysia. *International Journal of Economics and Finance, 6*(3), 126–138.

Ameer, R., & Othman, R. (2012). Sustainability practices and corporate financial performance: A study based on the top global corporations. *Journal of Business Ethics, 108*(1), 61–79.

Amini, M., & Bienstock, C. C. (2014). Corporate sustainability: An integrative definition and framework to evaluate corporate practice and guide academic research. *Journal of Cleaner Production, 76,* 12–19.

Amran, A., Zain, M. M., Sulaiman, M., Sarker, T., & Ooi, S. K., (2013). Empowering society for better corporate social responsibility (CSR): The case of Malaysia. *Kajian Malaysia: Journal of Malaysian Studies*, *31*(1).

Arrow, K. J. (1973). Social responsibility and economic efficiency. *Public Policy, 21*(3), 303–317.

Asif, M., Searcy, C., Zutshi, A., & Fisscher, O. A. (2013). An integrated management systems approach to corporate social responsibility. *Journal of Cleaner Production, 56,* 7–17.

Bowen, H. R. (2013). *Social responsibilities of the businessman*. Iowa: University of Iowa Press.

Bowen, H. R., & Johnson, F. E. (1953). *Social responsibility of the businessman*. New York: Harper Collins Publishing.

Brønn, P. S., & Vrioni, A. B. (2001). Corporate social responsibility and cause-related marketing: An overview. *International Journal of Advertising, 20*(2), 207–222.

Bruch, F. W. H. (2005). The keys to rethinking corporate philanthropy. *MIT Sloan Management Review, 47*(1), 49.

Carroll, A. B. (1991). The pyramid of corporate social responsibility: Toward the moral management of organizational stakeholders. *Business Horizons, 34*(4), 39–48.

Carroll, A. B., & Shabana, K. M. (2010). The business case for corporate social responsibility: A review of concepts, research and practice. *International Journal of Management Reviews, 12*(1), 85–105.

Carroll, A. B. (1979). A three-dimensional conceptual model of corporate performance. *Academy of Management Review, 4*(4), 497–505.

Carroll, A. B. (1983). Corporate social responsibility: Will industry respond to cutbacks in social program funding. *Vital Speeches of the day, 49*(19), 604–608.

Carroll, A. B. (1999). Corporate social responsibility: Evolution of a definitional construct. *Business and Society, 38*(3), 268–295.

Castka, P., Bamber, C. J., Bamber, D. J., & Sharp, J. M. (2004). Integrating corporate social responsibility (CSR) into ISO management systems–in search of a feasible CSR management system framework. *The TQM magazine, 16*(3), 216–224.

Dahl, R. A. (1957). The concept of power. *Behavioral Science, 2*(3), 201–215.

Darus, F., Hamzah, E. A. C. K., & Yusoff, H. (2013). CSR web reporting: The influence of ownership structure and mimetic isomorphism. *Procedia Economics and Finance, 7,* 236–242.

Foran, T. (2001). *Corporate social responsibility at nine multinational electronics firms in Thailand: A preliminary analysis*. Berkeley, CA: Nautilus Institute.

Friedman, M. (1970). A Friedman doctrine: The social responsibility of business is to increase its profits. *The New York Times Magazine, 13* (1970), 32–33.

Fuzi, N. M., Desa, A. F. N. C., Hibadullah, S. N., Zamri, F. I. M., & Habidin, N. F. (2012). Corporate social responsibility practices (CSR) and CSR performance in Malaysian automotive industry. *International Journal of Accounting and Financial Reporting, 2*(2), 268.

Geva, A. (2008). Three models of corporate social responsibility: Interrelationships between theory, research, and practice. *Business and Society Review, 113*(1), 1–41.

Givel, M. (2007). Motivation of chemical industry social responsibility through Responsible Care. *Health Policy, 81*(1), 85–92.

Gonzalez, M. C., & Martinez, C. V. (2004). Fostering corporate social responsibility through public initiative: From the EU to the Spanish case. *Journal of Business Ethics, 55*(3), 275–293.

Gordon, M., Lockwood, M., Vanclay, F., Hanson, D., & Schirmer, J. (2012). Divergent stakeholder views of corporate social responsibility in the Australian forest plantation sector. *Journal of Environmental Management, 113,* 390–398.

Haniffa, R. M., & Cooke, T. E. (2005). The impact of culture and governance on corporate social reporting. *Journal of Accounting and Public Policy, 24*(5), 391–430.

Hart, S. L. (1995). A natural-resource-based view of the firm. *Academy of Management Review, 20*(4), 986–1014.

Hassan, N. S., Lee, K. E., Mokhtar, M., & Goh, C. T. (2018a). The implementation of and reporting of corporate social responsibility in Malaysian chemical industries. *Asia Pacific Environmental and Occupational Health Journal, 4*(1), 1–7.

Hassan, N. S., Lee, K. E., Mokhtar, M., & Goh, C. T. (2018b). Mapping corporate social responsibility of chemical industries in Malaysia towards sustainable development: A conceptual framework. *Journal of Food, Agriculture and Environment, 16*(2), 215–220.

Hilson, G. (2012). Corporate social responsibility in the extractive industries: Experiences from developing countries. *Resources Policy, 37*(2), 131–137.

Hoq, M. Z., Saleh, M., Zubayer, M., & Mahmud, K. T. (2010). The effect of CSR disclosure on institutional ownership. *Pakistan Journal of Commerce and Social Sciences (PJCSS), 4*(1), 22–39.

Hove, H. (2004). Critiquing sustainable development: A meaningful way of mediating the development impasse? *Undercurrent, 1*(1).

Isa, S. M. (2012). Corporate social responsibility: What can we learn from the stakeholders? *Procedia-Social and Behavioral Sciences, 65,* 327–337.

Jones, T. M. (1980). Corporate social responsibility revisited, redefined. *California Management Review, 22*(3), 59–67.

Lantos, G. P. (2001). The boundaries of strategic corporate social responsibility. *Journal of consumer marketing, 18*(7), 595–632.

Lee, K. E., Mokhtar, M., Goh, C. T., Singh, H., & Chan, P. W. (2015). Initiatives and challenges of a chemical industries council in a developing country: The case of Malaysia. *Journal of Cleaner Production, 86,* 417–423.

Lindgreen, A., & Swaen, V. (2005). Corporate citizenship: Let not relationship marketing escape the management toolbox. *Corporate Reputation Review, 7*(4), 346–363.

Lougee, B., & Wallace, J. (2008). The corporate social responsibility (CSR) trend. *Journal of Applied Corporate Finance, 20*(1), 96–108.

Lu, J. Y., & Castka, P. (2009). Corporate social responsibility in Malaysia–experts' views and perspectives. *Corporate Social Responsibility and Environmental Management, 16*(3), 146–154.

Luo, X., & Bhattacharya, C. B. (2006). Corporate social responsibility, customer satisfaction, and market value. *Journal of marketing, 70* (4), 1–18.

Matten, D., & Moon, J. (2004). Corporate social responsibility. *Journal of Business Ethics, 54*(4), 323–337.

Mattera, M., Baena, V., & Cerviño, J. (2012). Analyzing social responsibility as a driver of firm's brand awareness. *Procedia-Social and Behavioral Sciences, 58,* 1121–1130.

Mawhinney, M. (2008). *Sustainable development: Understanding the green debates*. Wiley.

McMurray, A. J., Islam, M. M., Siwar, C., & Fien, J. (2014). Sustainable procurement in Malaysian organizations: Practices, barriers and opportunities. *Journal of Purchasing and Supply Management, 20*(3), 195–207.

McWilliams, A., & Siegel, D. (2001). Corporate social responsibility: A theory of the firm perspective. *Academy of Management Review, 26*(1), 117–127.

Mitchell, R. K., Agle, B. R., & Wood, D. J. (1997). Toward a theory of stakeholder identification and salience: Defining the principle of who and what really counts. *Academy of Management Review, 22* (4), 853–886.

Mohamad, A. B. (2005). Undang-undang dan pengurusan alam sekitar menurut Islam. *Malaysian Journal of Environmental Management, 6*(2005), 107–124.

Mohd Radzi, N. A., Lee, K. E., Abdul Halim, S., & Siwar, C., 2018a. How realistic is 'good citizenship' of the corporate company? The CSR implementation among FORBES-listed top oil and gas companies. *International Information Institute (Tokyo). Information, 21* (4), 1333–137.

Mohd Radzi, N. A., Lee, K. E., Abdul Halim, S., & Siwar, C. (2018b). What drives them to do CSR? Another empirical study of CSR motives from the perspective of the internal and external stakeholders. *International Information Institute (Tokyo). Information, 21* (3), 909–928.

Mohd Radzi, N. A., Lee, K. E., Abdul Halim, S., & Siwar, C. (2018c). Integrative approach for corporate social responsibility: A case from the banking industry. *International Journal of Academic Research in Business and Social Sciences, 8*(3), 251–266.

Mohd Radzi, N. A., Lee, K. E., Abdul Halim, S., & Siwar, C. (2018d). An empirical study of critical success factors and challenges in corporate social responsibility (CSR) implementation: The case of selected corporate foundations in Malaysia. *International Journal of Academic Research in Business and Social Sciences, 8*(3), 70–90.

Moon, J. (2007). The contribution of corporate social responsibility to sustainable development. *Sustainable Development, 15*(5), 296–306.

Moon, J., Crane, A., & Matten, D. (2005). Can corporations be citizens? Corporate citizenship as a metaphor for business participation in society. *Business Ethics Quarterly, 15*(3), 429–453.

Mustafa, S. A., Othman, A. R., & Perumal, S. (2012). Corporate social responsibility and company performance in the Malaysian context. *Procedia-Social and Behavioral Sciences, 65,* 897–905.

Naser, K., Al-Khatib, K., & Karbhari, Y. (2002). Empirical evidence on the depth of corporate information disclosure in developing countries: The case of Jordan. *International Journal of Commerce and Management, 12*(3/4), 122–155.

Park, B. I., Chidlow, A., & Choi, J. (2014). Corporate social responsibility: Stakeholders influence on MNEs' activities. *International Business Review, 23*(5), 966–980.

Pinney, C. (2001). Imagine speaks out. How to manage corporate social responsibility and reputation in a global marketplace: The challenge for Canadian business.

Poitras, G. (1994). Shareholder wealth maximization, business ethics and social responsibility. *Journal of Business Ethics, 13*(2), 125–134.

Porter, M. E., & Kramer, M. R. (2006). The link between competitive advantage and corporate social responsibility. *Harvard Business Review, 84*(12), 78–92.

Prathaban, V., & Rahim, N. (2005). Big earners, small givers. *Malaysian Business, 16,* 13–19.

Reidenbach, R. E., & Robin, D. P. (1991). A conceptual model of corporate moral development. *Journal of Business Ethics, 10*(4), 273–284.

Roberts, R. W. (1992). Determinants of corporate social responsibility disclosure: An application of stakeholder theory. *Accounting, Organizations and Society, 17*(6), 595–612.

Rooney, C. (1993). Economics of pollution prevention: How waste reduction pays. *Pollution Prevention Review*, *261*.

Saleh, M., Zulkifli, N., & Muhamad, R. (2010). Corporate social responsibility disclosure and its relation on institutional ownership: Evidence from public listed companies in Malaysia. *Managerial Auditing Journal, 25*(6), 591–613.

Schwartz, M. S., & Carroll, A. B. (2003). Corporate social responsibility: A three-domain approach. *Business Ethics Quarterly, 13*(4), 503–530.

Sharma, B. (2013). *Contextualising CSR in Asia: Corporate social responsibility in Asian economies.*

Siwar, C., & Harizan, S. H. M. (2009). *A study on corporate social responsibility practices amongst business organizations in Malaysia.* Bangi: Institute for Environment and Development, Universiti Kebangsaan Malaysia.

Smith, N. C., & Quelch, J. A. (1993). Ethics in marketing (Irwin, Homewood, IL). In *Socio-economic & Environmental Research Institute (SERI) (2008), Economic Briefings to the Penang State Government, SERI, Penang.*

Strand, R. (1983). A systems paradigm of organizational adaptations to the social environment. *Academy of Management Review, 8*(1), 90–96.

Suchman, M. C. (1995). Managing legitimacy: Strategic and institutional approaches. *Academy of Management Review, 20*(3), 571–610.

Sumiani, Y., Haslinda, Y., & Lehman, G. (2007). Environmental reporting in a developing country: A case study on status and implementation in Malaysia. *Journal of Cleaner Production, 15* (10), 895–901.

Székely, F., & Knirsch, M. (2005). Responsible leadership and corporate social responsibility: Metrics for sustainable performance. *European Management Journal, 23*(6), 628–647.

Tayşir, E. A., & Pazarcık, Y. (2013). Business ethics, social responsibility and corporate governance: Does the strategic management field really care about these concepts? *Procedia-Social and Behavioral Sciences, 99,* 294–303.

Teoh, H. Y., & Thong, G. (1984). Another look at corporate social responsibility and reporting: An empirical study in a developing country. *Accounting, Organizations and Society, 9*(2), 189–206.

Tsoutsoura, M. (2004). Corporate social responsibility and financial performance. UC Berkeley Working Paper Series.

Van Marrewijk, M. (2003). Concepts and definitions of CSR and corporate sustainability: Between agency and communion. *Journal of Business Ethics, 44*(2–3), 95–105.

Velasquez, M. G., & Velazquez, M. (2002). *Business ethics: Concepts and cases* (Vol. 111). Upper Saddle River, NJ: Prentice Hall.

Waworuntu, S. R., Wantah, M. D., & Rusmanto, T. (2014). CSR and financial performance analysis: Evidence from top ASEAN listed companies. *Procedia-Social and Behavioral Sciences, 164,* 493–500.

WBCSD. (1999). *Corporate social responsibility: Meeting changing expectations*. World Business Council for Sustainable Development.

Weber, M. (2008). The business case for corporate social responsibility: A company-level measurement approach for CSR. *European Management Journal, 26*(4), 247–261.

Witt, L. A. (1991). Exchange ideology as a moderator of job attitudes-organizational citizenship behaviors relationships 1. *Journal of Applied Social Psychology, 21*(18), 1490–1501.

Yam, S. (2012, January). Corporate social responsibility and the Malaysian property industry. In *18th Annual PRRES Conference, Adelaide, Australia* (pp. 15–18).

Zhao, Z. Y., Zhao, X. J., Davidson, K., & Zuo, J. (2012). A corporate social responsibility indicator system for construction enterprises. *Journal of Cleaner Production, 29,* 277–289.

# Shifting the Paradigm Toward Integrated Management of Urban River in a University Campus

Mohd Hafiyyan Mahmud, Khai Ern Lee, Mazlin Mokhtar, and Sharina Abdul Halim

**Abstract**

Urbanization has maximized the land use, causing natural rivers being concretized into urban rivers to accelerate excessive runoff as well as stabilize soil structure, which in turn expose to water pollution. Hence, urban river water is not seen as a valuable water resource. Alur Ilmu is an urban river in Universiti Kebangsaan Malaysia (UKM) that has been exposed to various sources of pollution in which existing management may not be effective in revitalizing and conserving the urban river. In this chapter, an integrated management framework has been proposed, consisting of structural and nonstructural approaches in revitalizing and conserving the water resource. Physical and biological treatments have been employed in structural approach to remove pollution at source, and this approach has successfully improved the water quality from Class II to Class III in less than a year, whereas initiatives in enhancing knowledge, attitude and practice as well as strengthening the participation of campus stakeholders have been employed in nonstructural approach. Combining structural and nonstructural approaches not only develops on-site treatment for the revitalization of the urban river, but also promotes social learning for the conservation of urban river. This integrated management framework is expected to shift the paradigm for the restoration and conservation of urban river to attain sustainability of water resource for the benefits of economic growth, social well-being and environmental protection.

**Keywords**

Urban river • Integrated management • Structural approach • Nonstructural approach • Water resource

M. H. Mahmud · K. E. Lee (✉) · M. Mokhtar · S. Abdul Halim
Institute for Environment and Development (LESTARI), Universiti Kebangsaan Malaysia, Bangi, 43600, Selangor, Malaysia
e-mail: khaiernlee@ukm.edu.my

## 1 Urban River

Urbanization increases the population working and living in urban areas as well as their socioeconomic activities. Hence, more areas are being developed to accommodate the needs of the population, such as roads, residential areas, public utilities, including rivers. Beginning in the twentieth century, urban population have increased more than 100% worldwide and reached 200% in less developed countries (Gupta 1984). In 2001, the world's largest city, New York, has a population of about eight million people; additionally, 17 cities have eight million inhabitants and Shanghai has more than 14 million inhabitants (United Nations 2004). According to the World Bank (2015), Malaysia is among the East Asian countries with fast growing development and population in urban areas; additionally, Malaysia was the fourth largest country in infrastructure development in the East Asia in 2010 with area development from 3900 to 4600 km$^2$ from 2000 to 2010.

As expected, urbanization is an inevitable process due to the development of infrastructure and socioeconomic activities to accommodate growing populations. Therefore, this effect has caused demands in making significant changes to the river system either directly or indirectly (Eyles 1997; Douglas 2005). These changes are permanent and may continue in the future; hence, it is feared that the water resource of the river in the urban area will be vulnerable to infrastructure development and socioeconomic activities along the river which could contribute to the pollution of the river, considering urbanization will have a significant impact on water drainage in terms of physical characteristics of river, socioeconomic functions and activities undertaken by stakeholders around the river.

Urban river is defined as a channel of water that has been concretized from either natural river or man-made river, located within urban area for the purposes of irrigation, water runoff as well as socioeconomic activities. Initially, the urbanization process set the conventional management


K. E. Lee (ed.), *Concepts and Approaches for Sustainability Management*, Advances in Science, Technology & Innovation, https://doi.org/10.1007/978-3-030-34568-6_4

paradigm of urban river for channeling excessive stormwater from upstream to downstream to avoid flood (Reese 2001; Zakaria et al. 2004; Mokthar et al. 2005). Due to the lack of environmental protection awareness, people tended to discharge all liquids including wastewater into the urban river, and this has led to pollution of water resources and caused detrimental effects to human and environment health. The focus of urbanization was given to economic development, and less emphasis was given to environmental protection because it was considered to be less cost-effective to the national economy (Thomas and Reese 2003); hence, it has resulted in poor management of urban river. The restoration and conservation of urban river within the post-development area are major challenges to conventional urban river management when attempting to radically shift the paradigm of urban river management to an alternative paradigm. In addition, efforts to conserve urban rivers within the post-development area require a high budget (Martin et al. 2007; ETP Annual Report 2012).

The restoration and conservation of urban rivers have gained the attention of developed and developing countries as rising demand from the public for better environmental health (EPA 2016), aesthetic appreciation and natural aesthetic gentrification, wishing to live in a healthy neighborhood, efficient and 'green' (Lim et al. 2013; Reese 2001) and improving the quality of life (Zakaria et al. 2004). In addition, the restoration and conservation efforts of urban rivers provide economic returns as well as potential as a source of local economic growth (ETP Annual Report 2012) and being a valuable water resource (Wong 2011; Lim and Lu 2016).

The evolution of urban river management in Malaysia is influenced by how the runoff and pollution control are determined based on the socioeconomic changes. It is undeniable that controlling the risks of flood for unusual heavy rain is the priority of the existing urban river management. This is due to the increase in the volume of running water as a result of intensive urbanization. However, the lack of consideration for pollution control and water quality will lead to chronic problems in the future. Hence, the balance of both functions in avoiding flood risk and controlling pollution is required in tandem with sustainable urban river management.

## 1.1 Features of Urban River Physical Landscape

Earth geomorphology is the physical state of the earth's surface, and it links to the urban geological structure and concrete water drainage to the surrounding area. It is influenced by runoff and discharge as well as human activities running at the riverside of the city (Toriman 2005). Factors affecting runoff and discharge are rainfall intensity, tree species in riparian areas, physical characteristics of river basins and land use in changing river structures (Wuriyati 2007). Physical features of the city's river and concrete water drainage are influenced by municipalities with paved surfaces. Due to the flow of urban rivers that are set in accordance with the norms set by humans, therefore the soil structure is more stable but it is prone to more runoff due to the surrounding paved surfaces (Nakamura et al. 2006) which carry pollutants into the river body (Din et al. 2012a, b) due to lack of canopy interception that allows water to infiltrate into the soil during rain events (Xiao et al. 1998). Due to surface pavement around the urban riverbank, it causes the lack of areas that can be used for aquatic and riparian habitats, therefore lack of flora and fauna species along the urban river and drainage (UKM 2013). These aforementioned factors result in changes of water flow patterns (hydrology), water quality of the river, exposure of urban river drainage system to runoff and land use from the surrounding areas that carry pollutants into the river body (Din 2012a, b). Thus, the functions and benefits of the urban rivers are still limited whereby it can only cater for irrigation purposes.

## 1.2 Functions of Urban River

The functions of rivers, urban rivers and concrete water drainages change according to respective purposes. Generally, urban rivers and concrete water drainages collect and channel runoff from the nearest water bodies, such as larger rivers, lakes, seas and dams (Mokthar et al. 2005; Department of Irrigation and Drainage Malaysia (DID) 2012; Chin 2006; Wohl and Merritts 2007; Speed et al. 2016) to avoid flooding. To date, most of the urban rivers and concrete water drainages maintain the same function and no other functions, such as recreation or as an alternative source of water, have been added. When the development of a city is the result of population growth, the function of urban rivers and the concrete water drainages also change and resemble a function like a natural river. Demands by the stakeholders within the city revolve around the gratification and appreciation of nature, the demand for safe, efficient and green areas (Reese 2001), improving the quality of life (Zakaria et al. 2004), demand for better environmental health (Environmental Protection Agency of Ireland 2011) and as a potential source of valuable resources that can give benefit directly and indirectly, such as harvesting runoff water and recharging underground water (Wong 2011, Lim and Lu 2016). Therefore, concrete water drainages, such as trenches or drains, are increasingly functioning and have a function and importance that are almost identical to the natural rivers. Therefore, the definition of concrete water drainage needs to be changed to urban river due to its increasing function.

### 1.3 Socioeconomic Activities of Stakeholders Around the Urban River

The human socioeconomic activities around urban rivers are dependent on urban river functions that are conceptualized by stakeholders. For example, Lim et al. (2013) found a socioeconomic change along the Cheonggyecheon River, South Korea, when it successfully conserved and preserved the river. Initially, the river was a concrete drainage under the highway. Due to the highway causing traffic congestion and air pollution to the nearby area, businesses in the area are experiencing difficulties due to the lack of visitors. But when the highway was demolished to conserve and preserve the water quality of the Cheonggyecheon River, the river was beautified, and the business in the area grew and became a tourist attraction for both locally and abroad. In this regard, the links between urban rivers and socioeconomic activities of stakeholders are very much related to each other.

### 1.4 Redefining Urban River

There are numerous studies related to the urban rivers (Douglas 1974; Douglas 1985; Wohl and Merritts 2007; Zakaria et al. 2004; United Nations 2018). However, the term 'water drainage' used by researchers is very loose whether the river, urban river and stormwater drainage or trench describing the same water drainage as having similar physical landscape features, functions, socioeconomic activities. Hence, the definition of urban river is still vague due to the existing definition inclining toward trench and drain systems (Gobster and Westphal 2004) or natural river (Wohl and Merritts 2007). Consequently, the definition of urban river is influenced by human perceptions and produces differences in function, water and environmental quality as well as social and economic development (Chin 2006; Gobster and Westphal 2004; Lim et al. 2013). The definition of an urban river is important as it determines whether the urban river should be conserved and preserved or otherwise (Gobster and Westphal 2004).

The definition of natural and modified rivers (by humans and natural events) is still debatable by researchers to date as the physical characteristics of the two types of rivers are almost identical (Wohl and Merritts 2007). This is because the effect of indirect modifications by the socioeconomic activities of the earlier civilization still preserves the nature of the river until to date and the effect of the river stabilization process also takes a very long time (Chin 2006; Wohl and Merritts 2007). However, concretization is a direct river modification that can be distinguished from natural river because of the different physical landscape features that do not have or maintain the characteristics expected for a natural river (Wohl and Merritts 2007).

Urbanization leads to the development of infrastructure and the vibrancy of economic activities in a given area leading to the construction of concretized rivers. The features seen from the stormwater drainage, trench, drain and urban river are almost identical. Its permanent construction works to channel runoff water rapidly to avoid flooding. However, many studies have shown that urban river is different from the drainage system in terms of function, physical landscape features and socioeconomic activities of the surrounding because humans receive many benefits from the urban river directly and indirectly other than channeling runoff water rapidly to avoid flooding.

As a result, urban river is defined as a river that has been concretized from either natural flow or man-made flow and it is located within urban area for the purpose of irrigation, water runoff as well as socioeconomic activities within the area. Therefore, water pollution resulted from urbanization has limited the functions and services of urban rivers in providing services for human and environmental purposes. In this regard, the sources and effects of urban river water pollution need to be identified to conserve and improve the functions of urban river.

## 2 Pollution and Pollution Control

In pursuing economic growth and urbanization, issues and problems related to the urban river environment are getting worse. Factors, such as aging of building and permanent infrastructure, rapid land clearing for infrastructure development and climate change, have put pressure on urban rivers, resulting in water quality deterioration (Keong 2006; Marsalek and Schreier 2009; University of British Columbia 2014). This is due to urban river being exposed to point-source and nonpoint-source pollutions.

Point-source pollutions are any sources of pollution that can be identified from which the pollutants are released, such as pipes, drains, boats or refineries (Hill 1997). The pollutants are produced by human activities, contributing to the deterioration of water quality of urban rivers and lakes (Thomas and Reese 2003). Nonpoint-source pollutions mean pollutions arising from soil runoff, precipitation, atmospheric deposition, drainage, seepage or hydrological changes. Nonpoint source pollutions, unlike pollutions from industrial and sewage treatment plants, where it comes from mixed sources. Nonpoint-source pollutions are caused by rain or runoff that move on the surface of the pavement and do not infiltrate into soil, but bring along pollutants which will eventually enter into lakes, rivers, wetlands, nearby coastal waters (Brezonik and Stadelmann 2002; Marsalek et al. 2008; United States Environmental Protection Agency 2018).

Water resources including urban rivers in Malaysia are exposed to point-source and nonpoint-source pollutions. Point-source pollution occurs in the urban river of Malaysia when the waste is released directly into the urban river. Domestic sources are usually derived from slaughterhouses, wet food shops, animal farms, household waste, cottage industry, domestic sewerage and agriculture. Malaysia has a tropical monsoon climate that is receiving hot and rain seasons only throughout the year. Due to the abundance of rainwater throughout the year, rivers including urban rivers in Malaysia face deterioration in water quality due to floods, flash floods, droughts, deposition of rivers due to soil erosion, sedimentation and solid waste. The root causes of nonpoint-source pollutions in urban rivers of Malaysia are from logging, land clearing for agriculture and development, sand mining, river reserve invasion, squatters and garbage collection centers (Keong 2006).

Point-source and nonpoint-source pollutions in urban river water not only affect water quality as a resource for human, but riparian and aquatic life is also affected. The effects of pollutions from sedimentation, organic matter, inorganic matter, nutrients, solid wastes and microorganisms have an impact on urban rivers and surrounding areas. Sedimentation has an impact on the quantity and quality of urban river water when suspended solid concentration is too high, causing turbidity to water especially in equatorial countries that receive high intensity of rain and hot weather throughout the year (Douglas 1974; Din et al. 2012a, b). When an urban river is concretized and receiving sedimentation throughout the year, it affects the depth of the urban river as a result of sedimentation (Chin 2006). Due to the deposition of suspended solids, sediments, soils and so on in the urban river, it causes urban river to become shallow. In addition, sedimentation causes urban rivers to lose their aesthetic and recreational values and alter aquatic and riparian habitats of urban river (Environmental Protection Agency of Ireland 2011).

In addition, pollutants from inorganic matters are highly toxic chemicals and can affect the health of humans and other organisms if they consume it. It affects the reduction of aquatic species, such as algae, invertebrates and fish communities causing riparian degradation (Paul and Meyer 2001; Meyer and Wallace 2001; World Bank 2006; Zhou et al. 2012). Additionally, inorganic substances, such as pesticides, and other highly toxic substances tend to accumulate in aquatic life and go into human and animal feedstock if they are controlled properly (Carson 2002; Atlanta Regional Commission 2002). Organic matters and high nutrients make an ideal condition for algae to bloom leading to eutrophication and reduction of dissolved oxygen, making it difficult for aquatic organisms' respiration processes to take place (Colangelo and Jones 2005; Speed et al. 2016). Moreover, the lack of dissolved oxygen will cause an anaerobic environment for the decomposition of organic matters and this gives an unpleasant odor to urban rivers (Atlanta Regional Commission 2002). Solid wastes and microorganisms entered into urban river water bodies will result in the loss of aesthetic and recreational value as well as causing riparian habitat degradation (Atlanta Regional Commission 2002) and vulnerability to various dangerous diseases when using water for drinking and recreational purposes (Thomas and Reese 2003).

To ensure the sustainability of the quantity and quality of urban river water and its environment to remain clean and healthy, pollution controls must be included in urban river management to prevent any form of pollution from entering urban river water bodies as well as removing pollutants in the water bodies. Best management practices are introduced to prevent any form of pollutions from entering nearby urban river water bodies, whereby best management practices apply control either physical or cultural functioning individually or as a group, in line with the source, location and climate of the area (United States Environmental Protection Agency 1993; Marsh 2011) for the restoration and conservation of urban rivers. Integrated management practices are divided into two parts: structural and nonstructural approaches (United States Environmental Protection Agency 1993). Structural approach focuses on reducing sources of pollution that have entered urban river by using technical capacity that includes scientific analysis and design of engineering systems to identify pollution sources and apply water treatment system as close as possible to the sources of pollution. Meanwhile, nonstructural approach focuses on preventing and controlling pollution before entering urban river (Thomas and Reese 2003) through changing stakeholders' attitudes toward environmental protection. Integration of both structural and nonstructural approaches to urban river management is expected to provide a paradigm shift that can link environmental protection, social empowerment and economic benefits to urban communities.

## 3 Current Urban River Management

Various studies have shown that the evolution of urban river management is reactive which relies on short-term solution to current problems and is not proactive (Reese 2001; Martin et al. 2007). Proactive solutions examine current issues, anticipate future problems and generate integrated solutions to address those problems. For the restoration and conservation of urban rivers for improved water quality, existing urban river management requires a paradigm shift toward a holistic and integrated approach due to the complexity of interrelated problems.

Pre-development paradigm that has less emphasis on environmental protection in urban river development planning has led to water quality deterioration. Therefore, current urban river management requires a new approach to restore and conserve urban river's water quality. The evolution of urban river management includes natural rivers, drainage systems, stormwater drainages and drains that are highly impacted by human practices from either urbanization or socioeconomic activities in the surrounding areas. In general, residents and urban planners still consider that stormwater channels are the final destination of waste disposal and affect the quality of human health, environment and security (Department of Irrigation and Drainage 2012; Lim et al. 2013). Hence, it is extremely difficult for the restoration and conservation of the urban environment when urban planning fails to incorporate environmental care elements, especially in urban river water quality.

Presently, the restoration and conservation of urban river are different from the previous management paradigm. According to Speed et al. (2016), the challenge of restoration and conservation of urban river is to balance between the natural functions of urban river and specific human needs. Furthermore, the complexity and scale of the restoration and conservation project lead to the failure to resolve pollution issues because of failing to take into account the processes at the basin level. Operations at a large scale require issues, consideration and participation of various stakeholders as well as planning and management tools. As such, it increases the uncertainties like climate change, land use, population growth and urban development toward the challenging future conditions to ensure that urban river is suitable for restoration.

The issues faced by urban river management are the best approaches to achieving the objectives, approaches and constraints of implementation, operation and maintenance, the diversity of issues to be addressed, the best way to conserve and the external problems resulting in the effectiveness of treatment. Aspects seen in the selection of the best approaches to achieving the objectives are the effectiveness of the water treatment system in the selected field (National Audit Department of Malaysia 2017). As funding has been spent on the development of high-cost water treatment system (National Audit Department of Malaysia 2017), it is desirable for the system to treat urban river effectively and the period of treatment does not take long to restore urban river's water quality (Chan 1999; Weng et al. 2003; Department of Irrigation and Drainage Malaysia 2012). In addition, there are constraints in the implementation of water treatment system whereby the feasibility of the treatment system varies depending on the condition of urban river and this complicates the installation of on-site treatment systems due to inadequate treatment system specifications (National Audit Department of Malaysia 2017; Thomas and Reese 2003). When water treatment system in the field has been financed, it is probable that it will not return the cost and value returned from the water treatment also takes a long time (Weng et al. 2003).

Operation and maintenance play an important role in the sustainability of water treatment in the field where costs are needed to repair water treatment system in the field to operate optimally, as a result of aging, damage caused by vandalism or clogging (Marsalek and Schreier 2009; Martin et al. 2013). In addition, urban river conservation needs to address various issues as it requires a comprehensive treatment scale, whereby if it is not sufficient, that water treatment system cannot improve water quality to a better level (Marsalek and Schreier 2009). In addition, external problems can affect the effectiveness of water treatment especially during flash floods and landslides that originate from natural disasters or anthropogenic incidents happened at the upstream are often ignored in urban river conservation (Marsalek 2003).

In addition to the issues faced to conserve urban rivers, the process of maintaining urban rivers also has its own unique aspects and issues. Among the issues faced are budget issues, socioeconomic and institutional mandates, defined returns from urban river care, inadequate planning, inefficient in enforcement and changes in stakeholders' practices. Urban river conservation issues such as budget, socioeconomic and institutional mandates as well as defined returns from urban river care are due to constraints and balancing interests and demands. This is because master plan planning and analysis use high cost (Marsalek and Schreier 2009; Chan 1997; Lim et al. 2013). In addition, the discomfort of squatters should be taken into account as they need to get out of their own homes when the government directs their relocation to another area. Hence, balancing stakeholders' needs and environmental care needs to be done fairly and equitably.

Planning failure is a matter of concern when there is less value in the formation of urban river master plan. Additionally, inefficient acts and enforcements as well as changes in stakeholders' practices are major issues when the failure of enforcement in regulating various acts to prevent pollution (Meenakshi and Mageswari 2002). Changes in stakeholders' practices have no apparent value because the practice changes based on small-scaled programs have less significant impact and it is difficult to evaluate their effectiveness (Weng et al. 2003).

Similarly, the management of river basins has undergone a change from the main purpose and use. Reyhan (2013) has summarized the management of river basins from the 1970s to the 1990s. Beginning in the 1970s, basin management aims to protect the infrastructure of local areas and resources at the downstream. This reflects the structural equation of the paradigm of river basin and urban river management that stated by Reese (1991). As stated by Mokthar et al. (2005), the use of river basin management is to stabilize the soil structure to prevent landslides. To address current problems, in the 1980s, the management has changed to an approach other than engineering and collaborates among other organizations. This change was made to manage resources, such as land, water and plants sustainably. However, cooperation between other organizations is difficult to build and technical approaches clearly fail to resolve the problems. In the 1990s, river basin management took into account the conservation of resources as well as enhanced the lives of local stakeholders by identifying and implementing integrated interventions and using approaches in prioritizing the involvement of local stakeholders in addressing local river basin problems with the help of technical methods. Management that integrates both science and social approaches is seen to be able to solve the problem of managing the local river basin and improve the lives of stakeholders that depend on the river basin. Therefore, the implementation of integrated management is necessary to manage urban river.

## 4 Integrated Management of Urban River

The Dublin Principles have stated in its first principle that freshwater is a limited and endangered source, essential for survival, economic development and the environment (Principles 1992). The population of the world increased by a factor of three during the twentieth century but the use of water resources increased by a factor of seven. It is estimated that one-third of the world's population reside in countries experiencing moderate to high water stress. This ratio is expected to increase to two-third by 2025 (Global Water Partnership 2000). In Malaysia, urban migration is a new challenge and complicates the formation of integrated management plans to incorporate elements of social unity and create economic opportunities. Furthermore, demand for domestic water supply is increasing and is expected to reach 16,176 million liters/day by 2050 (Keong 2006). No matter where the freshwater sources come from, either from rivers, lakes, dams and seas, these precious water resources need to be seen as a basin (Reese 2001). Consequently, the sustainability of water resources depends on how it is managed and water resources should be managed holistically and integrated by involving all levels of society.

Integrated water resource management (IWRM) (Global Water Partnership 2000) is an approach that promotes the coordination of development and management of water resources, land and related resources, to maximize economic and social welfare in a fair manner without affecting ecosystem conservation. IWRM covers the linkages between water resource managements including freshwater and saltwater, groundwater and surrounding life as well as the relationship between human and environmental systems. The concept of IWRM has been widely debated, and existing definition for the use of water bodies in cities, such as urban rivers, is still unclear. Therefore, regional and national institutions must develop the definition of their own IWRM practices that adopt the Global Water Partnership (GWP) 2000 framework. IWRM uses the best management practices in water resource management. For the restoration and conservation of urban river, IWRM is based on pollution control at its source. Any water received in the area should be controlled using best management practices and not directly releasing into urban river. Best management practices can be classified into two approaches, namely structural and nonstructural approaches (Micheal et al. 2004), whereby best management practices' goal is to provide pollution control from point-source and nonpoint-source pollutions to comply with the standards and guidelines set by the authorities (Thomas and Reese 2003).

Previous researches on the restoration and conservation of urban rivers in post-development areas emphasize integrated and holistic water management in addressing local problems and have their own perspective goals. However, contemporary research frameworks are complex, extensive and unique to each other, to support the institutionalization of urban river management according to their social and economic demographics and geomorphology (Martin et al. 2007; Al Bakri et al 2008; Wong 2011). It challenges decision-makers, engineers, urban river basin managers, local communities and stakeholders to build and implement their own approach based on existing research because broad principles need to be considered and uncertainties need to be taken into account and still lacking or do not have multi-discipline integration for solutions to complex problems, for example, feasibility study of effective water treatment system, stakeholder's participation and economic benefits of urban river management (Martin et al. 2007; Barbosa et al. 2012). Moreover, these approaches are specific to individual local issues, for example, stormwater management for industrial sectors (Wong et al. 2002), water harvesting as a water resource (Yang and Cui 2012), etc. However, it provides an unclear picture of the holistic main principle and must be included in the restoration and conservation of urban river. It is undeniable that various factors, such as restoration and conservation goals, types of approaches used and constraints as well as uncertainties, arise during the course of restoration and conservation processes of the urban river. Therefore, the central approach is to prevent pollutants from entering the urban rivers' water bodies. In this regard, decision-makers, engineers, river basin managers, local communities and stakeholders are required to empower integrated management to shift urban rivers that have only one function toward the restoration and conservation in the best possible way for the environment to be properly maintained as well as adding value to the socioeconomic status of locals.

## 4.1 Structural Approach

For the restoration of urban river water quality, structural approach is one of the best management practices to remove pollutants from water bodies (Lim and Lu 2016; Thomas and Reese 2003). Structural approach focuses on the removal of pollutants that have entered the urban river and provides quality and quantity control of water using technical capacity that includes scientific analysis and engineering-designed systems to identify sources of pollution and to apply the water treatment systems closest to the source. These include interpolation analysis (Murphy et al. 2009) and extrapolation (Wong et al. 2002) of water quality monitoring data for the spatial distribution of pollution sources.

Similar to many developed countries, the Malaysian government has outlined the National Water Quality Standard for freshwater, namely the Water Quality Index to identify and classify the level of water quality acceptance for human consumption and environmental health (Department of Environment 2006). The Water Quality Index consists of six water quality parameters, namely dissolved oxygen (DO), pH, ammoniacal nitrogen ($NH_3$–N), total suspended solids (TSS), chemical oxygen demand (COD) and biological oxygen demand (BOD). The value of the six parameters will be calculated according to the subindex calculations together with the weightage for each parameter to get a value that combines all subindex values. Of these Water Quality Index values, it will be classified into five classes and their usage whereby the highest value will be classified as Class I, which shows clean water and is most suitable for human consumption. The increase in the Water Quality Index Class shows the water quality is more polluted based on the six parameters, which can be detrimental to human and the environment health.

After the sources have been identified, field treatment will be carried out, for example, permeable pavement walls, gross waste traps, constructed wetlands, etc (Department of Irrigation and Drainage Malaysia 2012). One of the ways in placing the field treatment is using treatment train that is a series of retrofitted field treatment systems (United States Environmental Protection Agency 2000) without changing the geomorphological state to remove certain pollution in the body of water. The treatment efficiency relies on the ability of field treatment systems to address the targeted water quality parameters in the urban river. Design criteria for the installation and development of water treatment system as well as field operations depend on the size and mechanisms involved. Water treatment specifications are a requirement for optimum efficiency and better performance for the suitability of water treatment management (Lim and Lu 2016; Thomas and Reese 2003; Murphy et al. 2009). Hence, structural approach is designed in a set of action plans using scientific analysis that optimizes both site's characteristics and selected water treatment systems by reducing pollution from its sources according to standards set by local authorities or governments.

## 4.2 Nonstructural Approach

Another approach to improving the quality of urban river water is through nonstructural approach, whereby this approach prevents and regulates pollutants from entering the urban river water body besides removing pollutants released into the urban river water body (Thomas and Reese 2003). Nonstructural approach focuses on changing the behavior of stakeholders or the practices that cause urban river pollution using social capacity. This approach includes social analysis and social convention to identify and understand the practices that cause pollution and apply appropriate intervention on the stakeholders in accordance with the area, thereby preventing any subsequent pollution to the urban river (Bartlett 2005; Sparkman and Walton 2017).

Nonstructural approach is divided into two stages, namely assessing the management paradigm of urban river and applying social convention. Assessing the current urban river management paradigm is to identify the level of knowledge, attitudes in reflecting the practices as well as participation of stakeholders in the restoration and conservation of urban river and assess their perceptions in water quality and quantity (Gobster and Westphal 2004; Bartlett 2005). This assessment requires a set of questionnaires and a series of consultations with urban river stakeholders to understand the current management paradigm of the urban river, aiming to dissolve the complexity of nonpoint-source pollution by identifying practices and challenges for change as well as methods to address the local problems.

Prerequisite to the aforementioned methods, social convention is the establishment of intervention on current practices and understandings of environmental protection and nurturing salient practices in the restoration and conservation of urban rivers by enhancing the social well-being of stakeholders and adding value to the economy that they are relied on (Taylors and Wong 2002; Sparkman and Walton 2017). It includes pollution control, establishing advisory committees, education as well as capacity building, regulatory and practice development designed to limit the conversion of rainwater to runoff (Thomas and Reese 2003; Martin et al. 2007). The baseline program focuses on joint planning and pollution management through the participation and capacity building of stakeholders involved. This includes educating the public on the disposal of solid wastes in a cleaner and efficient way, legal regulations for waste disposal, change in material usage, work practice change, substitution of materials, reconstruction of drainage into the urban river by diverting into separate water treatment.

## 4.3 Challenges of Structural and Nonstructural Approaches

The challenge of structural approach in best management practices is the combination of complex stormwater and runoff water contributing to various types of biological and chemical reactive pollution in the urban river. Despite careful planning of water treatment systems in the field, pollution reduction is still inconsistent (Thomas and Reese 2003; Brown and Clarke 2007; Lim and Lu 2016). Therefore, it is difficult to achieve replicated results in treatment efficiency because structural approach can only reduce identified point-source pollutions. However, urban rivers are still vulnerable to nonpoint-source pollutions that enter into the water body. The greater volume of pollutants, such as nutrients, heavy metals and toxic substances enter into the urban river, the more efficient and robust water treatment system in the field is required. As such, the cost of water treatment will be higher for long-term operation and maintenance.

Figure 1 shows structural and nonstructural approaches in improving urban river water quality (Mahmud et al. 2017a). In a hypothetical situation, conventional urban river management is limited to functions of which irrigation and flood control cannot prevent and control pollution that are contributing to the deterioration of water quality by anthropogenic activities. It is because urban rivers receive pollutants from the upstream caused by natural events, economic development and anthropogenic activities that are difficult to control as it is beyond the capacity of conventional urban river management. Structural approach is designed to allow urbanization and economic development

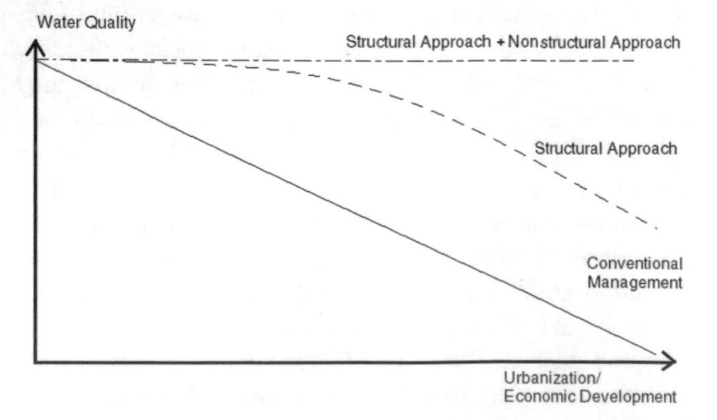


**Fig. 1** Hypothetical water quality comparison of urban river using conventional management, structural approach and nonstructural approach (Mahmud et al. 2017a)

by reducing its impact on the environment (Urbonas 2001; Thomas and Reese 2003). With the rapid urbanization and economic development, anthropogenic activities contribute to the diversity of pollutants, the water treatment capacities in the field are limited to the amount of pollutants discharged and increasing pollutant load will cause water treatment in the field less effective. Besides that, urban rivers are still vulnerable to external factors and problems, such as oil spills or flood events that carry pollutants from nonpoint sources, such as sediment, silts and others from upstream catchment. Furthermore, the effectiveness of water treatment in the field is based on the types of pollutants found in the urban river water body. Additionally, urban rivers have slow recovery process in water quality due to several factors, namely slow exchange of chemical composition between the water and soil due to the concrete base (Chin 2006; Wohl and Merritts 2007). Hence, relying only on structural approach is unsustainable because water treatment systems in the field can only treat point-source pollution, whereas nonstructural approach encourages behavior change in stakeholders' consent and practice toward environmental protection whereby it prevents and controls all forms of pollutions before entering the urban river water body instead of removing pollution from the water body (Thomas and Reese 2003). It requires capacity building, regulation and time to change the behavior of which they have been practicing so far. Therefore, both structural and nonstructural approaches are needed in integrated management of urban river.

## 4.4 Concept

Integrated management of urban river is a complex program that combines both structural and nonstructural approaches to current management in supporting a wide range of disciplines, including ecology, aquatic biology, hydrology and hydraulics, geomorphology, engineering, planning, communication, economics and social sciences to solve the problems of urban river. The current urban river management is on the verge of revolutionary change in response to the growing demand for water resources in urban areas due to the rapid economic development and human activities (Global Water Partnership 2000; Bahri 2012). Current water resources management needs to be resilient to climate change, competitions, conflicts, deficiencies and pollution of water resources; therefore, rethinking the concept of conventional urban river management is important in order to shift the paradigm of managing urban water cycle separately to integrated water resource management in urban areas supported by all stakeholders (Global Water Partnership 2000; Bahri 2012).

Integrated urban river management is adopting the concept of IWRM which contributes to the security of water resources in river basins within urban areas or as water supplies for rural areas. In addition, this integrated management leads to the opening up of the diversity of urban river potential from the economic, social and environmental perspectives by coordinating urban rivers from the aspects of water resources, gentrification, value-added economies and more. Additionally, it can unite various entities and groups of stakeholders around the urban rivers while solving the problems to achieve the goals of economic, social and environmental sustainability.

There are four important factors in the concept of integrated urban river management, namely (1) structural approach, (2) nonstructural approach, (3) understanding the paradigm (Reese 1991) and (4) the capacity for paradigm shift (Wong 2011). Implementing this concept requires the current urban river management paradigm and the capacity for paradigm shift. It is important to understand the relationship between physical problems, the effects of pollution and social problems that become the source of pollution where the problem of water quality requires both structural and nonstructural approaches to form integrated management of urban river. A social science platform is required to form integrated management of urban river, whereby it includes restoration and conservation of urban river for paradigm shift in urban river management. The concept should begin simultaneously for both structural and nonstructural approaches, whereby it focuses on mutually complementary solutions to address the urban river pollution within a time frame for effective outcomes (Lim and Lu 2016). Both structural and nonstructural approaches complement each other on the socio-science platform in which through this platform, demonstration of successful structural approach, it builds stakeholders' confidence in accountable and credible science in helping to build capacity and encourage more stakeholders' involvement throughout the process (Wong 2011). It enables the change to occur with confidence through the technological propagation and stimulates the emergence of development in a socio-technical environment for the paradigm shift of urban river management. Guidelines, collaborative goals, master plans and water quality standards provide a basis of vision and direction for future transition scenarios and routes for sustainable urban river management.

## 5 Case Study: *Alur Ilmu*, the National University of Malaysia

Urbanization contributes to population growth by improving physical development and socioeconomic development of the population and building rural areas to cities. These changes also change geomorphological structures including rivers in the area to strengthening soil integrity to nearby infrastructure and channeling runoff during storm events to prevent flash floods. Sustainable urbanization processes emphasize environmental protection including the care of the quantity and quality of urban rivers and the surrounding environment for urban sustainability.

*Alur Ilmu* was originally a 1.79-km-long natural river that has been converted into an urban river, flowing through the main campus of the National University of Malaysia, Selangor, Malaysia, before flowing into the Langat River, one of the UNESCO HELP (Hydrology for the Environment, Life & Policy) River Basins. As shown in Fig. 2, *Alur Ilmu* receives water source from the Permanent Reserved Forest, stormwater and runoff water from paved surfaces, such as paved roads or cemented roads around the urban river. *Alur Ilmu* is surrounded by buildings and infrastructures, comprising faculties, residential colleges, tar and pedestrian areas, administrative buildings and other paved areas which are full of socioeconomic activities of campus stakeholders. *Alur Ilmu* works to strengthen the buildings nearby soil structure and serves as an irrigation system to drain excess stormwater in the event of rain to prevent flooding. However, its initial development of less emphasis on the environmental protection has caused *Alur Ilmu* to be exposed to point-source and nonpoint-source pollutions along the river (UKM 1979). In addition, the current management and programs that have been implemented are less effective in restoring and conserving the water quality of the urban river. Consequently, *Alur Ilmu* records a decline in water quality as a result of water pollution (Chong 1999; Mokthar et al. 2005; Din et al. 2012a, b) and now such water quality makes the urban river neither regarded as a valuable water resource nor used for recreational purposes because of its low aesthetic value. According to the studies of the Water Quality Index, the water quality has decreased from Class I before the construction of the campus to Class II in 1999 (Chong 1999), Class III in 2003 (Mokthar et al. 2005), Class IV in 2012 (Din et al. 2012a, b), and Water Quality Index is expected to decline further if no action taken as shown in Fig. 3.

*Alur Ilmu* is exposed to water pollution arising from natural events and nearby anthropogenic activities. During rain, runoff water flows and collects pollutants, such as sediment, organic matter, solid waste, oil and grease from the pavement and into the urban river. As a result, it contributes to the deterioration of the Water Quality Index. Although the upper Ghazali Lake has sediment trap, it has failed to prevent sediment to flow into *Alur Ilmu* during heavy rains. This is because the sediment trap has surpassed its capacity to cope with sediment load as a result of erosion brought by runoff water. In addition, nearby anthropogenic activities, such as the cafeteria and faculties, also discharge wastewater into the water body contributing to pollution (Din et al. 2012a, b).

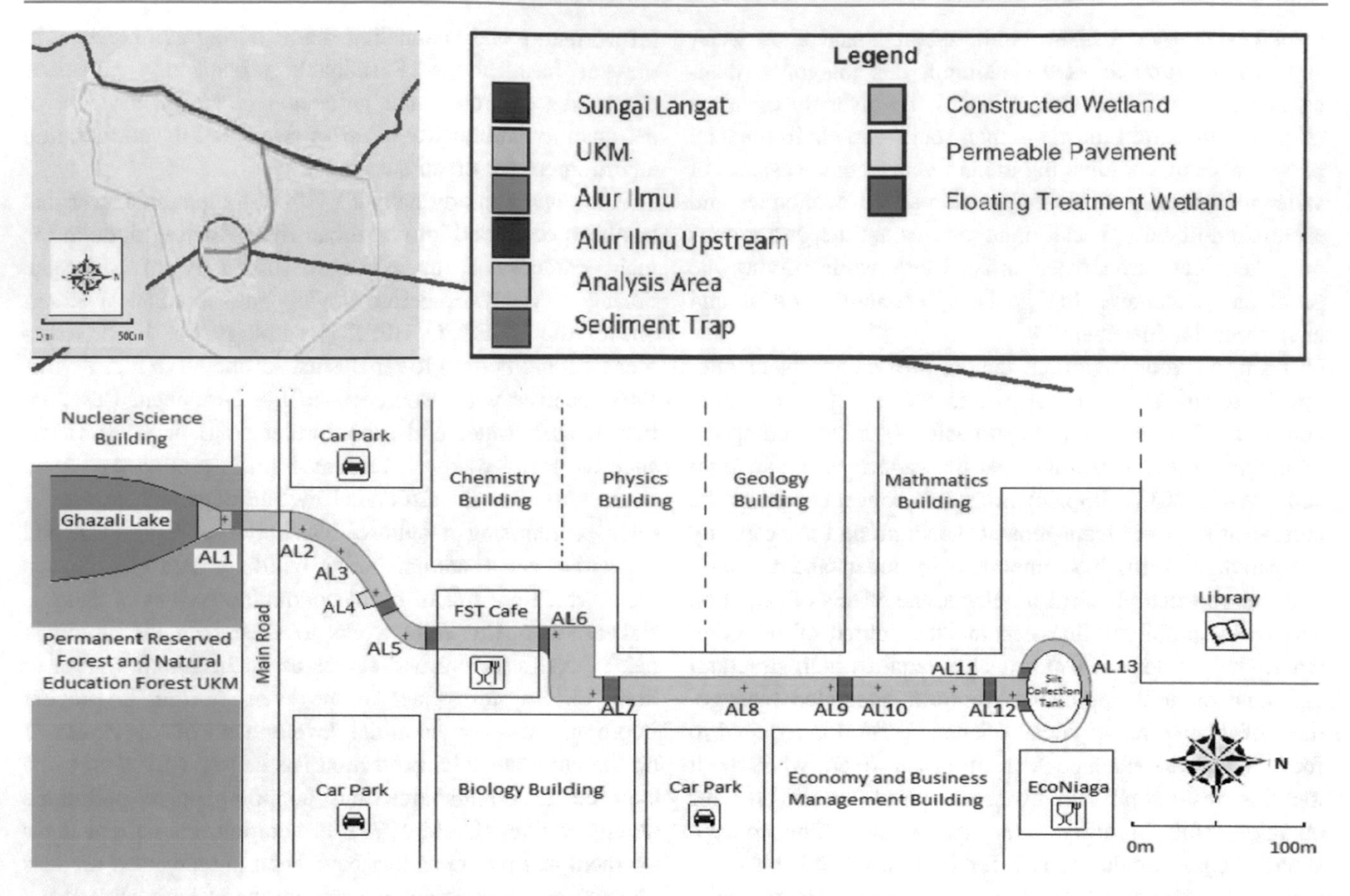


**Fig. 2** Geographical location of *Alur Ilmu* within the National University of Malaysia campus

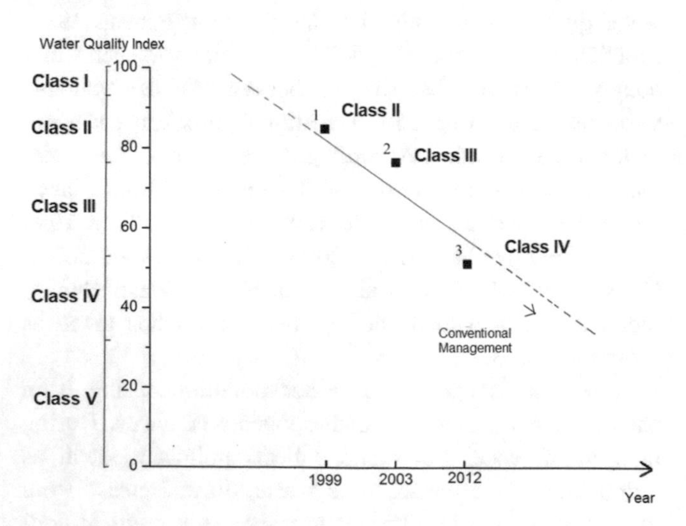


**Fig. 3** Water Quality Index of *Alur Ilmu* from 1999 to 2012. [1]Chong (1999); [2]Mokthar et al. (2005); [3]Din et al. (2012a, b)

In addition, the knowledge of stakeholders in protecting urban river and the attitude toward the management practices play an important role in influencing the water quality as the amount of pollutants from point source and nonpoint sources entering into *Alur Ilmu* depends on the management practices in restoring and conserving the urban river. If the conventional management of *Alur Ilmu* continues, pollution brought by the campus could result in the degradation of the ecosystem and affecting the water quality of Langat River which is one of the local water resources. Consequently, *Alur Ilmu* requires an integrated management to restore and conserve its water resource. The paradigm shift toward sustainable management of urban river is expected to restore the water quality by adopting a structural approach, and the health and the quality of life in the campus can be addressed by adopting a non-structural approach. Hence, best management practices that integrate both structural and nonstructural approaches are needed for integrated management of urban river.

## 6 Shifting the Paradigm of Urban River Management

Figure 4 shows the integrated management framework for urban river focusing on *Alur Ilmu*. The integrated framework for urban river is divided into three stages in three different platforms (i.e., science, social and social science): Stage 1 co-framing the problem collaboratively; Stage 2 co-producing knowledge-based and transferable solutions by establishing integrated management of urban river; and Stage 3

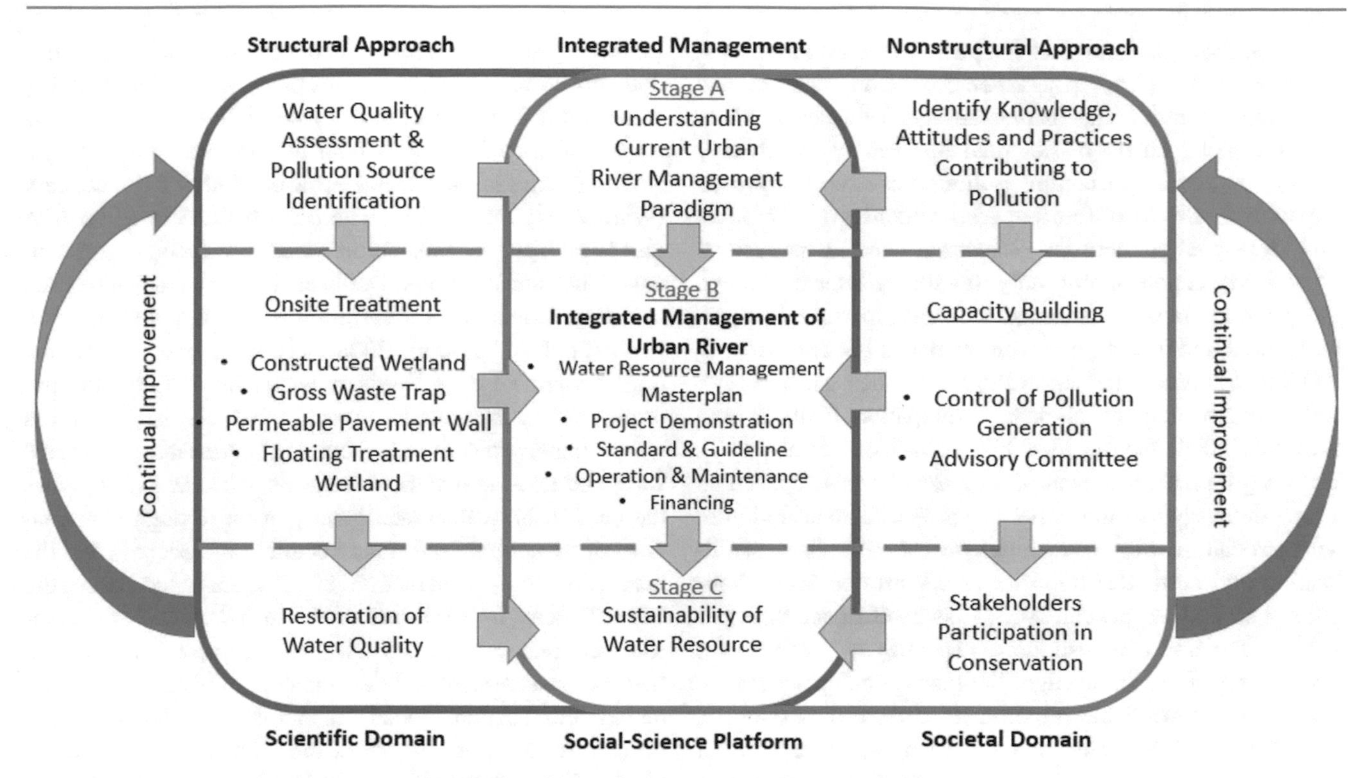


**Fig. 4** Integrated management framework for urban river focusing on *Alur Ilmu* (Lee et al. 2018)

co-implementing the knowledge-based solutions generated by both scientific and societal domains.

(i) Stage 1: Co-framing the problem collaboratively

The first stage of integrated management of urban river begins with co-framing the problem collaboratively for *Alur Ilmu* among the scientific and societal domains which consist of related multi-disciplines and multi-stakeholders with experience, expertise or any other aspects that are relevant to the problem (Pohl and Hadorn 2008). This group of stakeholders will form a committee to understand current urban river management paradigm in *Alur Ilmu* by conducting water quality assessment and pollution source identification through structural approach by scientific domain, and identifying knowledge, attitudes and practices contributing to pollution through nonstructural approach by societal domain. Setting up this working committee with a structured organization is important where responsibility, efficiency and decision-making are clearly defined, whereby this is to establish a balance between the scientific domain (researchers) and societal domain (stakeholders) at every level to form a shared leadership for paradigm shift in urban river management (Scholz et al. 2006). Understanding the current urban river management paradigm of *Alur Ilmu* needs to define complex sustainability issue as a relevant social issue and raise questions for scientific research (Lang et al. 2012; Siebenhüner 2004; Wiek et al. 2012), and it will be instrumental for the formation of integrated management that balances the scientific and societal importance in knowing and solving the real problems of *Alur Ilmu*.

(ii) Stage 2: Co-producing knowledge-based and transferable solutions

In order to develop collaborative knowledge-based and transferable solutions for *Alur Ilmu* (Lang et al. 2012; Lee et al. 2018), the integration of concepts and findings throughout the collaborative research is essential to make pragmatic research a success. Therefore, the responsibility of each stakeholder at all different levels must be determined first as this is to implement transparent management taking into account inertia, reluctance and structural barriers (Maasen and Lieven 2006; Wiek 2007). There are different levels of engagement of stakeholders, including tasks, responsibilities, technical skills, levels of concern and willingness to contribute time and energy (Thomas and Reese 2003). In addition, to enable transdisciplinary integrated management, cognitive-related leadership (providing a way to integrate stakeholders' epistemic differences), structure (addressing the need for coordination and information exchange) and procedure (resolving conflicts during the process) are important elements that must be present in such management (Gray 2008).

Integrated management of *Alur Ilmu* requires the integration of structural and nonstructural approaches on the social science platform that includes water resource management master plan, project demonstration, standard and guideline, operation and maintenance and financing. Water resource management master plan that set by an advisory committee aiming to control pollution is able to give an impact not only to restore but also to conserve *Alur Ilmu*. The relationship between structural and nonstructural approaches conducts not only on-site treatment through gross waste traps, constructed wetlands, permeable pavement walls and floating treatment wetlands but also capacity building for enhancing knowledge, attitudes and practices and strengthening stakeholder participation (Gobster and Westphal 2004; Ison et al. 2007; Pahl-Wostl et al. 2008). The integrated management of *Alur Ilmu* has also opened up a new dimension to the impact of project demonstration in which social learning can impact on natural values, gentrification and aesthetics (Gobster and Westphal 2004; Marsalek et al. 2008; Lim et al. 2013; Viswanathan and Schirmer 2015). Project demonstration can be a catalyst for improvement of knowledge, attitudes and practices of stakeholders toward the importance of *Alur Ilmu* (Gobster and Westphal 2004; Lim and Lu 2016). As such, project demonstration is imperative to any river restoration and conservation efforts that are either proactive or reactive because they can trigger a change in stakeholder practices toward river restoration and conservation. Finally, integrated management of *Alur Ilmu* should set standard and guideline to control pollution to prevent any pollution to happen (Thomas and Reese 2003) and should be supported by operation and maintenance as well as financing to cover the costs of managing *Alur Ilmu*.

(iii) Stage 3: Co-implementing knowledge-based solutions generated by both scientific and societal domains

The final stage of integrated management of *Alur Ilmu* is to shift the management paradigm toward ensuring the sustainability of water resource through establishing the principles for the integration and application of knowledge-based solutions, namely (1) continual improvement based on the outcome, (2) generating science and social values, and (3) assessing science and social impacts. Continual improvement is reflected in the results obtained from structural and nonstructural approaches in managing *Alur Ilmu*. For structural approach, the review on the restoration of *Alur Ilmu* is based on the Water Quality Index and it is proposed with the addition of other water quality parameters, namely oil and grease and heavy metals, whereas for nonstructural approach, the review of the participation of stakeholders in the conservation of *Alur Ilmu* should be made on their knowledge, attitudes and practices toward pollution control and participation in management. Likewise, shared learning between science and social domains through project demonstration has a high visibility value and it is important to carry out different criteria in the review of contributions as both perspectives adhere to the quality of criteria, such as scientific credibility or practicality (Wiek 2007; Jahn 2008). The contributions to solve *Alur Ilmu*'s problem by stakeholders take into account both scientific and social aspects (Defila et al. 2006) and can be used as a transformation to scientific innovation and social progress (Pohl and Hadorn 2008). Figures 5 and 6 show the improvement of *Alur Ilmu*'s water quality before and after implementing structural approach that consists of gross waste traps, constructed wetlands, permeable pavement walls and floating treatment wetlands, whereas Fig. 7 shows the need to have integrated management through combination of structural and nonstructural approaches for the restoration and conservation of *Alur Ilmu* by improving Water Quality Index further. Believing *Alur Ilmu* as a water resource, integrated urban river management helps in sustainable management and development of water resource by taking into account social, economic and environmental benefits whereby it recognizes the differences and requirements of each stakeholder and entity that use or abuse water and environment (GWP and INBO 2009). Hence, uplifting *Alur Ilmu* as part of the IWRM will ensure the sustainability of water resource, not only for the use of campus stakeholders but also for environmental and habitat protection despite rapid urbanization and economic development.

## 7 Conclusions

The effects of urbanization on urban rivers have changed the geomorphological structure of the land by strengthening the integrity of the soil to the nearby infrastructure besides the runoff during storm events to prevent flash floods. In addition to its permanent structure of urban river which did not emphasize on environmental protection, the rapid growth of socioeconomic activities, the construction of buildings and the opening of the surrounding land have made urban river susceptible to various types and forms of pollution from point-source and nonpoint-source pollutions throughout the urban river. As a case study, *Alur Ilmu* of the National University of Malaysia has been adopted as an example of urban river in this study. Although many efforts have been undertaken to restore and conserve *Alur Ilmu*, the water quality has continued to decline to an alarming and unsustainable level. Hence, the current management of the urban river needs to be addressed in order to resolve the problem.

Integrated management of *Alur Ilmu* has been proposed, whereby it consists of structural and nonstructural approaches on the social science platform. Treatment taken place in

**Fig. 5** Comparison of Alur Ilmu before (top) and after (bottom) implementation of structural approach (Mahmud et al. 2017b, c, 2018)

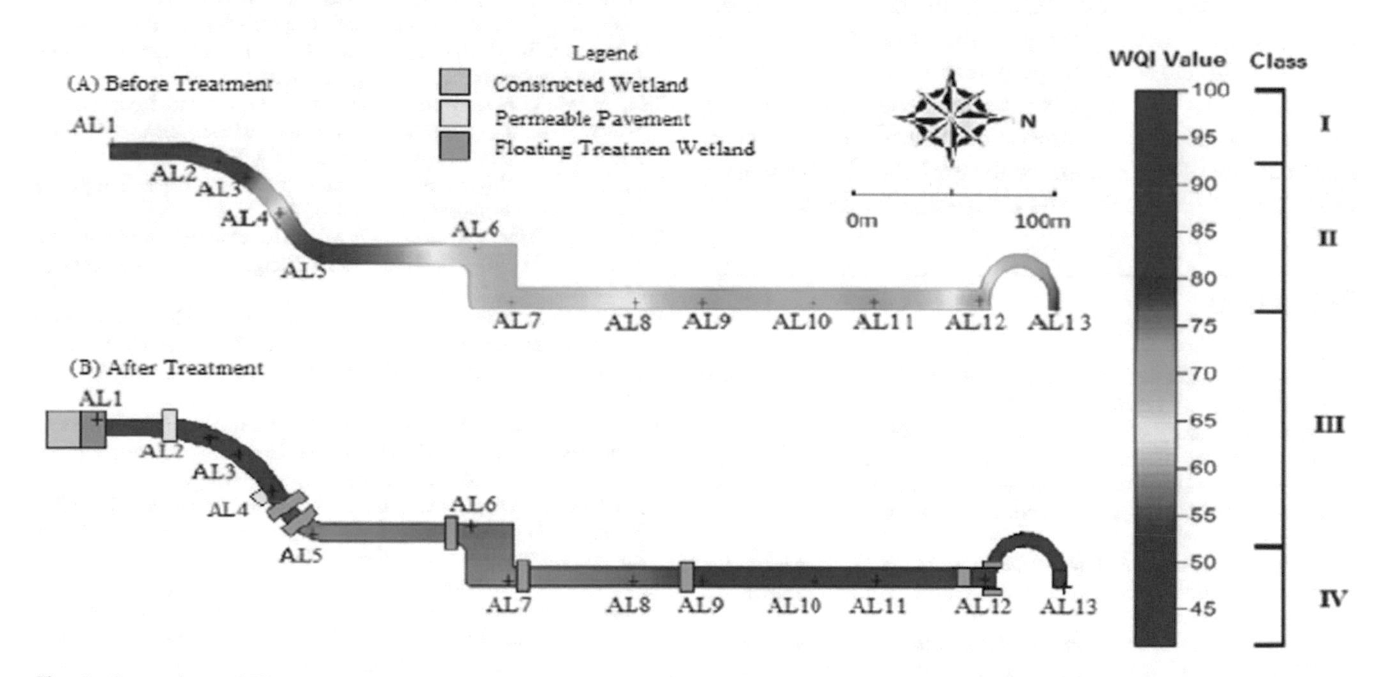


**Fig. 6** Comparison of Water Quality Index before and after the implementation of structural approach (Mahmud et al. 2017b, c, 2018)

the field through structural approach is gross waste traps, constructed wetlands, permeable pavement walls and floating treatment wetlands. Due to the location of *Alur Ilmu* is surrounded by buildings and infrastructure, hence most of the areas are paved and have very limited space. Therefore, selected water treatment infrastructures are of low-impact development without changing the physical landscape and geomorphology of the urban river. Structural approach taken has successfully improved from Class III to Class II in less than a year. Nonstructural approach taken focuses on enhancing knowledge, attitude and practice as well as strengthening the participation of campus stakeholders.

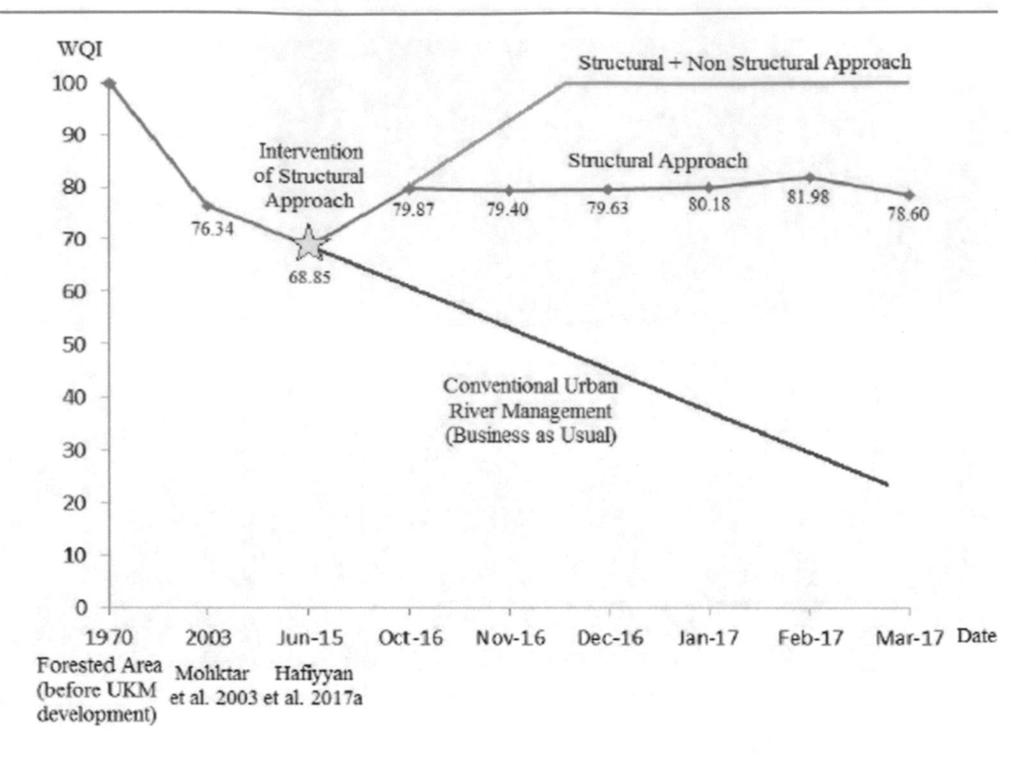


**Fig. 7** Water Quality Index of *Alur Ilmu* and hypothetical structural and nonstructural approaches in improving water quality

Combining structural and nonstructural approaches not only develops on-site treatment for the restoration of urban river, but also creates social learning for the conservation of urban river through stakeholders' participation to avoid pollution from occurring. This study is expected to be used as a model for the restoration and conservation of urban river to attain sustainability of water resource for the benefits of economic growth, social well-being and environmental protection.

## References


Al Bakri, D., Rahman, S., & Bowling, L. (2008). Sources and management of urban stormwater pollution in rural catchments, Australia. *Journal of Hydrology, 356*(3–4), 299–311.

Atlanta Regional Commission. (2002). *The georgia stormwater manual*. Georgia: Atlanta Regional Commission.

Bahri, A. (2012). Integrated urban water management. *TEC Background Papers, 16*.

Barbosa, A. E., Fernandes, J. N., & David, L. M. (2012). Key issues for sustainable urban stormwater management. *Water Research, 46* (20), 6787–6798.

Bartlett, C. (2005). *Stormwater knowledge, attitude and behaviors: A 2005 survey of North Carolina residents*. NC Department of Environment and Natural Resources.

Brezonik, P. L., & Stadelmann, T. H. (2002). Analysis and predictive models of stormwater runoff volumes, loads, and pollutant concentrations from watersheds in the Twin Cities metropolitan area, Minnesota. *United States of America. Water Research, 36*(7), 1743–1757.

Brown, R. R., & Clarke, J. M. (2007). *Transition to water sensitive urban design: The story of Melbourne, Australia* (Vol. 7). Melbourne: Facility for Advancing Water Biofiltration, Monash University.

Carson, R. (2002). *Silent spring*. Houghton Mifflin Harcourt.

Chan, N. W. (1997). Mitigating Natural Hazards and Disasters in Malaysia: Lessons Learnt from Tropical Storm Greg and its Aftermath. In *Proceedings of the First International Molaysian Studies Conference,* 11–13 August, Kuala Lumpur.

Chan, N. W. (1999). WaterConsen1otion, Reuse and Reduction of Water. Working paper in: *Sustainable Management of Water Resources in Malaysia Workshop*, 20 July, Kuala Lumpur.

Chin, A. (2006). Urban transformation of river landscapes in a global context. *Geomorphology, 79,* 460–487.

Chong, W. L. (1999). *Effluent study of column oxidation and drainage system within UKM*. Bachelor Thesis, Bangi: National University of Malaysia.

Colangelo, D. J., & Jones, B. L. (2005). Phase I of the Kissimmee river restoration project, Florida, USA: Impacts of construction on water quality. *Environmental Monitoring and Assessment, 102*(1–3), 139–158.

Defila, R., Di Giulio, A., & Scheuermann, M. (2006). Forschungsverbundmanagement: Handbuch für die Gestaltung inter-und transdisziplinärer Projekte. vdf Hochschulverlag AG.

Department of Environment. (2006). *Malaysia Environmental Quality Report 2006*. Kuala Lumpur: Ministry of Energy, Science, Technology, Environment and Climate Change Malaysia.

Department of Irrigation and Drainage Malaysia. (2012). *Urban stormwater management manual for Malaysia (MSMA)* (2nd ed.). Kuala Lumpur: Department of Irrigation and Drainage Malaysia.

Din, H. M. (2012). *Analysis of water quality and outflow time series at Alur Ilmu the National University of Malaysia*. M.A Dissertation, Bangi: National University of Malaysia.

Din, H. M., Toriman, M. E., Mokhtar, M., Elfithri, R., Aziz, N. A. A., Abdullah, N. M., et al. (2012a). Loading concentrations of pollutant in Alur Ilmu at UKM Bangi campus: Event mean concentration (EMC) approach. *The Malaysian Journal of Analytical Sciences, 16,* 353–365.

Din, H. M., Toriman, M. E., Mokhtar, M., Elfithri, R., Aziz, N. A. A., Abdullah, N. M., et al. (2012b). Loading concentrations of pollutant in Alur Ilmu at UKM Bangi campus: event mean concentration (EMC) approach. *Malaysian Journal of Analytical Sciences, 16*(3), 353–365.

Douglas, I. (1974). The impact of urbanization on river systems. *The International Geographical Union Regional Conference*.

Douglas, I. (1985). Urban sedimentology. *Progress in physical geography, 9*(2), 255–280.

Douglas, I. (2005). Urban geomorphology. In P. G. Fookes, E. M. Lee, & G. Milligan (Eds.), *Geomorphology for Engineers* (pp. 757–779). Boca Raton, FL: Whittles Publishing.

Environmental Protection Agency Ireland. (2011). *Parameters of water quality. Interpretation and Standards*. Wexford, Ireland: Environmental Protection Agency of Ireland.

ETP Annual Report. (2012). *NKEA: Greater Kuala Lumpur and Klang Valley*. Kuala Lumpur: ETP.

Eyles, N. (1997). Environmental geology of Urban areas. *Geological Canada, 21*(4), 590.

Global Water Partnership. (2000). *Integrated water resources management*. Sweden: Global Water Partnership.

Gobster, P. H., & Westphal, L. M. (2004). The human dimensions of urban greenways: Planning for recreation and related experiences. *Landscape and urban planning, 68*(2–3), 147–165.

Gray, B. (2008). Enhancing transdisciplinary research through collaborative leadership. *American Journal of Preventive Medicine, 35*(2), S124–S132.

Gupta, A. (1984). Urban hydrology and sedimentation in the humid tropics. In J. E. Costa & P. J. Fleisher (Eds.), *Developments and applications of geomorphology* (pp. 240–267). Berlin, Heidelberg: Springer.

GWP & INBO. (2009). *A handbook for integrated water resources management in basins*. Sweden: Elanders.

Hill, M. S. (1997). *Understanding environmental pollution*. United Kingdom: Cambridge University Press.

Ison, R., Röling, N., & Watson, D. (2007). Challenges to science and society in the sustainable management and use of water: Investigating the role of social learning. *Environmental Science & Policy, 10*(6), 499–511.

Jahn, T. (2008). Transdisciplinarity in the practice of research. In *Transdisziplinäre Forschung: Integrative Forschungsprozesse verstehen und bewerten. Campus Verlag, Frankfurt/Main, Germany*, pp. 21–37.

Keong, C. W. (2006). *River restoration in Malaysia*. Kuala Lumpur: Department of Irrigation and Drainage Malaysia.

Lang, D. J., Wiek, A., Bergmann, M., Stauffacher, M., Martens, P., Moll, P., et al. (2012). Transdisciplinary research in sustainability science: Practice, principles, and challenges. *Sustainability Science, 7*(1), 25–43.

Lee, K. E., Abdullah, R., Hanafiah, M. M., Halim, A. A., Mokhtar, M., Goh, C. T., et al. (2018). An integrated approach for stakeholder participation in watershed management. *Environmental risk analysis for asian-oriented, risk-based watershed management* (pp. 135–143). Singapore: Springer.

Lim, H., Kim, J., Potter, C., & Bae, W. (2013). Urban regeneration and gentrification: Land use impacts of the Cheonggye Stream Restoration Project on the Seoul's central business district. *Habitat International, 39*, 192–200.

Lim, H. S., & Lu, X. X. (2016). Sustainable urban stormwater management in the tropics: An evaluation of Singapore's ABC waters program. *Journal of Hydrology, 538*, 842–862.

Maasen, S., & Lieven, O. (2006). Transdisciplinarity: A new mode of governing science? *Science and Public Policy, 33*(6), 399–410.

Mahmud, M. H., Lee, K. E., Halim, S. A., Mokhtar, M., & Er, A. C. (2017a). Revitalization of Urban Stream through Integrated Stormwater Management: A Conceptual Framework for Structural and Non-Structural Approaches. *PROCEEDINGS ICOSH-UKM 2017*, 4–7 April, Malaysia, pp. 223–229.

Mahmud, M. H., Lee, K. E., & Mokhtar, M. (2017b). On-site phytoremediation applicability assessment in Alur Ilmu, Universiti Kebangsaan Malaysia based on spatial and pollution removal analyses. *Environmental Science and Pollution Research, 24*(29), 22873–22884.

Mahmud, M. H., Lee, K. E., Mokhtar, M., & Ab Wahid, M. (2018). Technical strategy for revitalizing urban river water quality through structural approach at National University of Malaysia (UKM) Bangi Campus, Selangor. *Malaysia. Applied Ecology and Environmental Research, 16*(3), 2681–2699.

Mahmud, M. H., Lee, K. E., Mokhtar, M., Ab Wahid, M., Goh, T. L., Norbert, S., et al. (2017c). Spatial distribution of water quality index in stormwater channel: a case study of Alur Ilmu, UKM Bangi Campus. *Asia Pacific Environmental and Occupational Health Journal, 3*(1), 33–38.

Marsalek, J. (2003). Road salts in urban stormwater: An emerging issue in stormwater management in cold climates. *Water Science and Technology, 48*(9), 61–70.

Marsalek, J., Jimenez-Cisneros, B., Karamouz, M., Malmquist, P.-A., Goldenfum, J., & Chocat, B. (2008). *Urban water cycle processes and interactions*. The Netherlands: Taylor & Francis.

Marsalek, J., & Schreier, H. (2009). Overview of the theme issue, innovation in stormwater management in Canada: the way forward. *Water Quality Research Journal of Canada, 44*(1).

Marsh, J. (2011). *BMPs & management measures structural & non-structural*. Pojoaque Pueblo: US EPA Region 5 NPS Program.

Martin, C., Ruperd, Y., & Legret, M. (2007). Urban stormwater drainage management: The development of a multicriteria decision aid approach for best management practices. *European Journal of Operational Research, 181*, 338–349.

Meenakshi, R., & Mageswari, S. (2002). Rivers: Issues and Problems in Relation to Communities. In: N. W. Chan (Ed.), *Rivers: Towards sustainable development* (pp. 443–450). Penang: Universiti Sains Malaysia Publisher.

Meyer, J. L., & Wallace, T. B. (2001). Lost linkages in lotic ecology: Rediscovering small streams. In: M. C. Press, N. J. Huntly, & S. Levin (Eds.), Ecology: achievement and challenge, (pp. 295–317). Boston: Blackwell Science.

Mokthar, M., Chee, F. H., Chong, W. L., Ooi, Y. Y., & Tan, B. H. (2005). Water quality study at Alur Ilmu, UKM Bangi Campus: Towards an integrated water resource management approach. *The Malaysian Journal of Analytical Sciences, 12*(3), 388–395.

Murphy, R. R., Curriero, F. C., & Ball, W. P. (2009). Comparison of spatial interpolation methods for water quality evaluation in the Chesapeake Bay. *Journal of Environmental Engineering, 136*(2), 160–171.

Nakamura, K., Tockner, K., & Amano, K. (2006). River and wetland restoration: lessons from Japan. *BioScience, 56*(5), 419–429.

National Audit Department of Malaysia. (2017). *Auditor-general report year 2017 on activities of federal/ministries/departments and federal statutory bodies, Series 1*. Kuala Lumpur: National Audit Department of Malaysia.

Pahl-Wostl, C., Mostert, E., & Tàbara, D. (2008). The growing importance of social learning in water resources management and sustainability science. *Ecology and Society, 13*(1), 24.

Paul, M. J., & Meyer, J. L. (2001). Streams in the urban landscape. *Annual Review of Ecology and Systematics, 32*, 333–365.

Pohl, C., & Hadorn, G. H. (2008). Core terms in transdisciplinary research. In H. G. Hadorn, H. Hoffman-Riem, S. Biber-Klemm, W. Grossenbacher-Mansuy, D. Joye, C. Pohl, U. Wiesmann, E. Zemp (Eds.), Handbook of transdisciplinary research (pp. 427–432). Dordrecht: Springer.

Principles, D. (1992, January). The Dublin statement on water and sustainable development. In *International conference on water and the environment* (pp. 26–31).

Reese, A. J. (1991). Successful municipal storm water management: key elements. In *Proceedings of the 15*th *annual conference of the association of floodplain managers* (pp. 202–205). Denver, Colorado.

Reese, A. J. (2001). Stormwater paradigms. *Stormwater Magazine.* July–August 2001.

Reyhan, E. (2013). Stakeholder involvement in sustainable watershed management. In: O. Murat (Ed.), *Advances in landscape architecture.* Rijeka: Intech, p. 281.

Scholz, R. W., Lang, D. J., Wiek, A., Walter, A. I., & Stauffacher, M. (2006). Transdisciplinary case studies as a means of sustainability learning: Historical framework and theory. *International Journal of Sustainability in Higher Education, 7*(3), 226–251.

Siebenhüner, B. (2004). Social learning and sustainability science: which role can stakeholder participation play. In: *Proceedings of the 2002 Berlin conference on the human dimension of human change knowledge for the sustainability transition. The Challenge for Social Science*, July 2004, pp. 76–86.

Sparkman, G., & Walton, G. M. (2017). Dynamic norms promote sustainable behavior, even if it is counternormative. *Psychological Science, 28*(11), 1663–1674.

Speed, R., Tickner, D., Naiman, R., Gang, L., Sayers, P., Yu, W., et al. (2016). *River restoration: a strategic approach to planning and management*. UNESCO Publishing.

Taylors, A., & Wong, T. (2002). *Non-structural stormwater quality best management practices—An overview of their use, value, cost and evaluation*. Victoria, Australia: Cooperative Research Centre for Catchment Hydrology.

Thomas, N. D., & Reese, A. J. (2003). *Municipal stormwater management*. United States of America: Lewis Publisher.

Toriman, M. E. (2005). Hydrology characteristics of urban river. In S. Mohamad, M. E. Toriman, K. Aiyub, & M. Jaafar (Eds.), *River and development*. Bangi: Universiti Kebangsaan Malaysia Press.

United Nations. (2004). Demographic Yearbook 2001. New York.

United Nations. (2018). *Sustainable cities: Why they matter.* Viewed 14 April 2018. https://www.un.org/sustainabledevelopment/wp-content/uploads/2016/08/11.pdf.

United States Environmental Protection Agency. (1993). United states environmental protection agency (USEPA) 1993. *NPDES best management practices manual*. USEPA: United States of America.

United States Environmental Protection Agency. (2000). *Low impact development (LID): A literature review*. Washington, D.C: Office of Water, Report EPA-841-B-00–005.

United States Environmental Protection Agency. (2016). *Achieving water quality through integrated municipal stormwater and wastewater plans.* Viewed 21 March 2017. https://www.epa.gov/sites/production/files/.../memointegratedmunicipalplans_0.

United States Environmental Protection Agency. (2018). Polluted runoff: Basic information about nonpoint source. Viewed 18 August 2018. https://www.epa.gov/nps/basic-information-about-nonpoint-source-nps-pollution.

Universiti Kebangsaan Malaysia. (1979). *Physical development report Batch 1978/79*. The National University of Malaysia, 4th Board Meeting.

Universiti Kebangsaan Malaysia. (2013). *UKM Bangi campus physical development masterplan 2007–2020* (2nd ed.). Bangi: Institute of Environment and Development.

University of British Columbia. (2014). *Integrated storm-water management Plan*. University of British Columbia (UBC).

Urbonas, B. (2001). Summary of emergent stormwater themes, american society of civil engineers. In *Linking stormwater BMP designs and performance to receiving water impact mitigation, Snowmass Village*, pp. 1–8.

Viswanathan, V. C., & Schirmer, M. (2015). Water quality deterioration as a driver for river restoration: A review of case studies from Asia, Europe and North America. *Environmental Earth Sciences, 74* (4), 3145–3158.

Weng, C. N., Abdullah, A. L., Ibrahim, Ab. L., & Ghazali, S. (2003). River pollution and restoration towards sustainable water resources management in Malaysia. In *Proceeding of National Conference on Society, Space and Environment in a Globalised World: Prospects & Challenges.*

Wiek, A. (2007). Challenges of transdisciplinary research as interactive knowledge generation–Experiences from transdisciplinary case study research. *GAIA-Ecological Perspectives for Science and Society, 16*(1), 52–57.

Wiek, A., Ness, B., Schweizer-Ries, P., Brand, F. S., & Farioli, F. (2012). From complex systems analysis to transformational change: A comparative appraisal of sustainability science projects. *Sustainability Science, 7*(1), 5–24.

Wohl, E., & Merritts, D. J. (2007). What is a natural river? *Geography Compass, 1*(4), 871–900.

Wong, T. H. F. (2011, September). Framework for stormwater quality management in Singapore. In *12th international conference on Urban drainage, Proceedings, Porto Alegre, Brazil, International Water Association.*

Wong, T. H., Fletcher, T. D., Duncan, H. P., Coleman, J. R. & Jenkins, G. A. (2002). A model for urban stormwater improvement: Conceptualization. In *Global solutions for Urban drainage* (pp. 1–14).

World Bank. (2006). Water resources management in Japan policy, Institutional and Legal Issues, World Bank Analytical and Advisory Assistance (AAA) Program. Viewed 5 February 2018. http://siteresources.worldbank.org/INTEAPREGTOPENVIRONMENT/Resources/.

World Bank. (2015). Malaysia among most urbanized countries in East Asia. Viewed 26 July 2017. http://www.worldbank.org/en/news/feature/2015/01/26/malaysia-among-most-urbanized-countries-in-east-asiaWRM_Japan_experience_EN.pdf.

Wuriyati, A. (2007). Review of geometry as terminology in river engineering research and development. *Journal of Water Resources Technology, 4*(3).

Xiao, Q., McPherson, E. G., Simpson, J. R., & Ustin, S. L. (1998). Rainfall interception by Sacramento's Urban forest. *Journal of Arboriculture, 24*(4), 235–244.

Yang, R., & Cui, B. (2012). Framework of integrated stormwater management of Jinan City, China. *Procedia Environmental Sciences, 13,* 2346–2352.

Zakaria, N. A., Ab gani, A., Abdullah, R., Sidek, L. M., Kassim, A. H. & Ainan, A. (2004). MSMA—A new urbans stormwater management manual for Malaysia. *Advance in Hydro—Science and Engineering,* 6. United States of America: The University of Mississippi. ISBN-13: 978-0937099124.

Zhou, T., Wu, J., & Peng, S. (2012). Assessing the effects of landscape pattern on river water quality at multiple scales: A case study of the Dongjiang River watershed, China. *Ecological Indicators, 23,* 166–175.

# Enculturing Sustainable Development Concept Through Chemistry Curriculum for Education for Sustainable Development

Suganty Kanapathy, Khai Ern Lee, Mazlin Mokhtar, Subarna Sivapalan, Sharifah Zarina Syed Zakaria, and Azizah Mohd Zahidi

**Abstract**

The enculturation of sustainable development concept through education system, especially via STEM subjects, is vital in developing citizens who are able to adopt sustainability as a life principle. Although STEM agenda has been emphasized in the national education blueprint, attention is needed on the role of STEM education especially chemistry subject toward education for sustainable development. In this chapter, a framework that encompassing six main areas, namely education policy, awareness, resources, curriculum orientation, pedagogical approach and stakeholder's engagement is proposed in guiding the enculturation of sustainable development concept through chemistry curriculum. The government and institutional policies play an important role to create awareness in understanding the concept and relevance of the sustainable development concept in chemistry subject. Training, finance and time have been identified as resources that need to be considered while integrating sustainable development concept in syllabus and contextualizing chemistry content that should be considered in the curriculum. In addition, the learning outcomes need to be revised based on the pedagogical types, learning approaches and pedagogical techniques used. Lastly, the framework will be instrumental for administrators and teachers to develop a sustainable chemistry teaching toward achieving education for sustainable development with the participation of stakeholders, including government, experts in the field of ESD, administrators and chemistry teachers, in enculturing sustainable development concepts through chemistry subject.

**Keywords**

Education for sustainable development • Chemistry • Sustainable development • STEM

## 1 Education for Sustainable Development

The concept of sustainable development was adopted at the United Nations General Assembly in 1987. Since the introduction of the concept of sustainable development, education approaches that support the sustainable development agenda have been explored. This is because education is recognized as the human right and foundation for a country's development where education today is essential to enhancing the ability of leaders and citizens to find new solutions and directions for a better and sustainable future. Generally, sustainable development requires change in terms of human thought and action (UNESCO 2013). Therefore, the approach through education is one of the important efforts in transforming society toward sustainable development (Foo 2013; Mahat et al. 2014).

Education for Sustainable Development (ESD) was given attention when world leaders agreed that the concept of sustainable development should be intensified as a global goal through education. ESD was formally recognized for the first time through the Earth Summit in Rio de Janeiro in 1992 known as Agenda 21. In Agenda 21 (1992) report, Chapter 36 specifically emphasized education in terms of: (1) improving basic education; (2) the adaptation of existing education toward sustainable development; (3) enhancing community understanding and awareness; and (4) training.

K. E. Lee (✉) · M. Mokhtar · S. Z. S. Zakaria
Institute for Environment and Development (LESTARI), Universiti Kebangsaan Malaysia, 43600 Bangi, Selangor, Malaysia
e-mail: khaiernlee@ukm.edu.my

S. Sivapalan
Centre for Social Transformation for Sustainable Lifestyles, Universiti Teknologi PETRONAS, 32610 Bandar Seri Iskandar, Perak, Malaysia

A. M. Zahidi
Pusat GENIUS@Pintar Negara, Universiti Kebangsaan Malaysia, 43600 Bangi, Selangor, Malaysia

S. Kanapathy
Pusat GENIUS@Pintar Negara, Universiti Kebangsaan Malaysia, 43600 Bangi, Selangor, Malaysia


K. E. Lee (ed.), *Concepts and Approaches for Sustainability Management*, Advances in Science, Technology & Innovation, https://doi.org/10.1007/978-3-030-34568-6_5

Therefore, practitioners have considered education as one of the key components to sustainable development. ESD is part of Agenda 21, covering environmental, social and economic dimensions to emphasize global issues. In addition, ESD is also defined as the integration of key sustainable development issues into the teaching and learning process (Crespo et al. 2017). According to Lenglet et al. (2010), ESD comprises a variety of concepts, theories, policy prescriptions and practical methods aimed at restructuring the education system with a particular focus on environmental, social and economic dimensions that are key dimensions for sustainable development. Considering the issue of sustainable development is complex, it is important to integrate these three dimensions to achieve sustainability successfully. ESD is an education that embraces the past, relevant to the present, and has a vision for the future (Pigozzi 2010).

Sustainable development has been a focus of Malaysia since the 1970s, with a special emphasis on eradicating poverty, improving people's well-being, providing universal access to education and caring for the environment. In 2009, Malaysia developed the New Economic Model (NEM) which identified three main goals of achieving high income, inclusiveness and sustainability, as the basis for Malaysia's development plan by 2020. In general, the NEM reflects the three dimensions of sustainable development, namely economic, social and environmental. At the same time, the Eleventh Malaysia Plan 2016–2020 (11MP) also forms the basis of the three dimensions. The theme of the 11MP is "Anchoring Growth on People" in which the people are central to all development efforts by ensuring that no community groups are left out to participate and benefit from national development. Critically, the 11 MP is in line with most of the Sustainable Development Goals (SDGs) (Economic Planning Unit 2015).

Malaysia has also given special attention to the ESD where Malaysia has adopted the principles of Agenda 21 as one of the most important sustainable development documents in the national planning process (Hassan 1998; Ngah et al. 2011). To promote ESD in Malaysia, six strategies have been implemented by the Ministry of Education at all levels of the education system in Malaysia. Among these strategies are: (1) expanding the supply of skilled and knowledgeable labor; (2) improving access to quality education and training; (3) improving the quality of education and training delivery system; (4) promoting lifelong learning; (5) increasing the supply of science and technology labor; and (6) reinforcing positive values (UNESCO 2011). In the UNESCO (2011) report, it is stated that ESD's priority in Malaysia is divided into three categories, namely environment, education and poverty reduction. Climate change is a new priority that the Ministry of Education is working to address through ESD.

At the same time, academic institutions have taken various initiatives to incorporate aspects of Agenda 21 into academic syllabus as well as campus-based activities. Studies have also shown that aspects of sustainability have been covered in science, social science and engineering at the school, graduate and postgraduate levels. In addition, aspects of sustainability have been introduced through various programs at school level, such as Technical and Vocational Education (Minghat and Yasin 2010) and Environmental Education (Said et al. 2007). However, students often fail to integrate sustainability concepts with their lessons (Reza 2016). For example, students from Ecology and Natural Resource Management have difficulty in integrating the knowledge gained with sustainability into sustainable urban infrastructure development or planning. This is because the aspect of sustainability is not emphasized in detail in the syllabus taught in Malaysia (Reza 2016).

## 2 STEM Education

Literacy in Science, Technology, Engineering and Mathematics (STEM) is an essential element of science-related programs in the twenty-first century. STEM education and research play a vital role in advancing technology, medicine, sustainability, agriculture, national security, economics and society, as well as finding answers to most questions related to human life. STEM education is a global concept that integrates critical, analytical, systematic and collaborative thinking processes in which students integrate their processes and concepts into the real world (Ndinechi and Okafor 2016).

Therefore, the vision of most STEM academic programs is to focus on workforce development and research to meet the needs of local and regional industries, national security and efforts to become more competitive in the global market (Egarievwe 2015). There are various definitions of STEM education that have been discussed based on different approaches, such as silo, implicit and integrated approaches. In the silo approach, each STEM subject is taught separately, and the focus is on the core subject. This approach uses a college-based method or high standard instruction, and students are expected to gain a deeper understanding of the course content. Implicit approach focuses on real-world problems in the social, cultural and contextual aspects of the knowledge domain. Students are exposed to problem solving and learning techniques through a variety of contexts. An integrated approach involves integrating STEM content to teach as a subject. This approach is divided into two types: multidisciplinary and interdisciplinary. Multidisciplinary enables students to associate knowledge domains with specific subjects, while through interdisciplinary, students can associate domains of knowledge and skills across subjects with critical thinking skills, problem solving and knowledge to solve real-world problems (Le et al. 2015). Each approach has its own strengths and weaknesses. Therefore, teacher's initiative is important in assessing the subject knowledge domain and selecting the best teaching approach to attract students toward STEM education.

Proposal to teach integrated STEM in the secondary curriculum has been encouraged in some countries, but it has recently become more and more important (Williams 2011). This integration is considered a solution to educational reform in the USA. This is because there is a need to provide a technologically qualified and sophisticated engineering workforce to develop a high-tech knowledge-based economy. However, the main focus of STEM education is to increase the competency of students in various disciplines to meet the needs of the workforce in the twenty-first century (Obama 2009). The lack of STEM workers that threaten the USA has demanded STEM education reform in the country and subsequently spread to many countries around the world (Marginson et al. 2013; Pitt 2009).

The Malaysian Education Blueprint was launched by the Ministry of Education for 2013–2025. In the blueprint, strengthening STEM has been identified as one of the key elements, whereby the main goal of the STEM initiative is to ensure that Malaysia has a sufficient number of STEM graduates to meet the industry needs that serve as the engine of national economy. This is important to achieve the national goal of the Malaysian government of producing one million STEM professionals by 2020. Therefore, various key criteria have been proposed to strengthen STEM delivery through the education system, including (1) increasing student interest through new learning approaches and strengthening the curriculum; (2) improving teachers' skills and abilities; and (3) increasing student and public awareness. STEM approaches refer to pedagogical strategies that emphasize the application of knowledge, skills and values from the disciplines of Science, Technology, Engineering and Mathematics to help students in solving real-world problems. As such, Malaysia sees education as one of the ways to achieve advanced countries to meet STEM-driven economic challenges and demands (Ministry of Education Malaysia 2013).

## 3 Knowledge, Attitude and Behavior of Sustainable Development

ESD serves as an "umbrella" for education and now has the potential to bring about change and innovation in education, teaching and learning. Various studies have been conducted to find out how knowledge of sustainable development concepts that can influence students' attitudes and behaviors.

### 3.1 Knowledge

Azapagic et al. (2005) conducted a study to determine knowledge of engineering students on sustainable development. The study was conducted on engineering students from 40 different universities from developed and developing countries. The results show that the level of student knowledge is modest, and there is a significant knowledge gap between the social and economic dimensions of sustainable development. However, students feel that sustainable development is more important to future generations than themselves.

Furthermore, Sheikh et al. (2012) studied the perceptions of first-year undergraduate engineering students on sustainable development at a public university in Malaysia, indicating that students are unaware of sustainable development and most of the students do not know how to explain the definition of sustainable development because it has never been exposed to the concept. Although students cannot explain the definition of sustainable development, most students are found to be involved in daily activities based on sustainable development, such as Earth Hour, recycling, green technology, and climate change. This study shows that students understand the importance of caring for the environment, but primary and secondary education that do not associate environmental dimensions with sustainable development causes a lack of knowledge about sustainable development.

### 3.2 Knowledge/Awareness and Attitude

Aziz et al. (2012) conducted a survey on first-year engineering students to develop a structural model to assess students' attitudes toward sustainability. The findings of the study show that the structural model produced is reliable and is compatible with the influence of knowledge on sustainability attitudes among first-year engineering students. In addition, the analysis also shows that basic knowledge of sustainable development is strongly linked to developing and enhancing student attitude. The model also measures the level of knowledge and attitudes about the need to understand the concept of sustainable development and changes students' attitudes toward sustainability.

At Plymouth University, Kagawa (2007) explored perceptions, understandings and attitudes on sustainable development and related concepts and issues. The study found that students relate the concept of sustainable development and sustainability to the environment in relation to economic and social dimensions. In addition, the study found that students have a common pro-sustainability attitude and express a variety of feelings about the future of society in facing sustainability-oriented challenges. In terms of changing themselves toward a more sustainable lifestyle, students are often exposed to responsible consumer behaviors, such as changing purchasing habits, recycling, saving energy and/or water and changing modes of transportation. Therefore, Kagawa (2007) suggested that students' learning more relevant to specific contexts is important for establishing a curriculum change process where student needs, aspirations and concerns for sustainability are addressed. These changes are expected to help students become more responsible citizens.

In addition, a study was also conducted to examine the change in attitude and perceptions of twelfth-grade students on chemical and environmental issues after studying environmental subject. The results show that students are experiencing significant changes in awareness of environmental issues as the subject is closely related to students' personal lives. It is suggested that increasing student awareness at the secondary level may affect students' attitudes toward environmental issues in the future. Researchers also believe that such programs can promote ESD concepts among students. However, curriculum changes in this study were found to be insufficient to induce significant differences in students' attitudes toward chemistry learning. Curriculum modification is necessary in order to have a more profound impact on changing student attitudes and to create a more environmentally conscious mind (Mandler et al. 2012).

## 3.3 Knowledge and Attitude

Said et al. (2007) conducted a study to find out students' understanding of the extent to which environmentally responsible behaviors and students' involvement in activities related to the nature. The results show that students' understanding of environmental concepts is relatively poor and that students are less clear in defining sustainable development concepts. However, the findings show that environmental education, both formal and informal, has increased students' awareness of the environment. However, environmental education has been found to be less effective in changing students' attitudes and behaviors. This study also suggests that by overcoming barriers, such as pedagogical approaches, changes in student behavior can be seen to facilitate a more sustainable lifestyle.

In addition, a study conducted by Meerah et al. (2010) primary, secondary and urban secondary school students in Malaysia prove that the use of environmental education as a medium to deliver sustainable development content in primary and secondary education in Malaysia. The results show that only 35% of school students have knowledge of environmental issues. However, the levels of knowledge and behavior are low among the students. Therefore, more emphasis should be placed on environmental education both inside and outside of the classroom. Although 35% is a low percentage of knowledge, this indicates that undergraduate students who have essentially completed primary and secondary education are exposed to environmental knowledge as one of the dimensions of sustainable development.

## 3.4 Knowledge, Attitude and Behavior

Idros (2006) discussed the levels of knowledge, attitudes and behaviors of university students in Malaysia toward the environment and its implications for education in explaining sustainable development. The results show that the students do not have a high level of knowledge of environmental facts but do show a better understanding of the concept of the environment. Students were also found to have a strong pro-environmental attitude despite the students' willingness to act' is less than the students' actual actions. This means that the strength of the students' behavioral patterns toward the environment is modest, indicating that there is a need to plan or reformulate the curriculum systematically with a focus on promoting a deeper understanding of environmental conditions. Students also need to be equipped with knowledge of action strategies, understanding the complexity of the problems involved, and more importantly, identifying the causes that lead to a less sustainable future.

Fah and Sirisena (2014) conducted a study to assess the level of knowledge, attitudes and behaviors among high school students (grade four) in urban and rural areas in Sabah, Malaysia. The analysis shows that grade four students do not show the expected results where the knowledge is the lowest compared to attitudes and behaviors. In general, the students' attitudes and behaviors toward the environment are high. It is also found that the low knowledge is due to less time spent by the Ministry of Education on environmental education, especially on some key topics in the environment. Fah and Sirisena (2014) emphasize that the purpose of environmental education in Malaysia is not achieved and that improvement is necessary to create awareness about the importance of environmental education in Malaysian secondary schools.

Gusti (2016) conducted a study in Indonesia on primary school students to assess the relationship between knowledge, attitudes and behaviors in sustainable waste management. The result shows that knowledge and attitudes toward sustainable waste management are significantly associated with sustainable waste management behaviors. Increasing students' interests or behaviors in sustainable waste management can be achieved by enhancing students' knowledge of sustainable waste management which includes reducing, reusing and recycling waste for more environmentally friendly energy production.

## 3.5 Knowledge and Awareness

Hassan et al. (2010) conducted a study to assess the level of environmental awareness based on the concept of sustainable development among high school students in the Hulu Langat District, Selangor, Malaysia. High school students show a high degree of awareness of the environment in relation to the concept of sustainable development. In addition, science students have a higher level of environmental awareness of the concept of sustainable development, indicating the need to learn more about environmental and ecological concepts resulting in higher knowledge of science students than those of literature students studying basic science where the syllabus is not sufficient.

Yuan and Zuo (2013) assessed students' level of awareness about sustainability and perceptions of students about higher education for sustainable development. It reveals that students generally have a high level of awareness of sustainability issues, but the level of students' interpretation of the meaning of higher education for sustainable development is low. In general, it is found that aspects of sustainability in the environment are given priority to students compared to other factors such as social responsibility.

Fabbrizzi et al. (2016) analyzed the level of knowledge and perceptions of the concepts of sustainability and well-being of high school students in Tuscany. The analysis reports that young people are less stressed about sustainability in schools where students' perceptions show that the concept of sustainability is related to environmental aspects. In relation to well-being, it shows how factors that can be associated with individual aspects are considered more important. The role of well-being indicators is important for policy making. Discussions about the perceptions and definitions of well-being indicators lead to forums where new ideas can be developed, and those indicators can be negotiated and agreed upon. Therefore, the introduction of key aspects of shaping the concept of well-being for the younger generation and student perceptions is the first step in providing concrete ways of representing the various dimensions.

## 4 Integration of Sustainable Development Concept

ESD is an area of interest and has been promoted through various United Nations agendas and initiatives, such as Agenda 21, Decade of Education for Sustainable Development 2005–2014 and Global Action Program by the United Nations Educational, Science and Cultural Organization (UNESCO) to implement ESD (Agenda 21 1992). However, ESD is often considered to be closely related to environmental education (Darwish et al. 2010; Hoang and Kato 2016). Johannesson et al. (2011) suggest that differentiating ESDs with environmental education in terms of content and teaching methods is a difficult task. Sterling (2001) suggests that ESD actually covers more topics than environmental education and can explain all education for the purpose of change. As such, much of the existing support for ESD is seen in many different areas, namely architectural education (Shari and Jaafar 2006), technical, entrepreneurship, vocational education and training (TEVET) (Gomani 2010), construction projects (Abidin 2010), Green ICT (Suryawanshi and Narkhede 2015), mathematics (Sivapalan 2016) and many others.

The Earth Summit held in Rio de Janeiro in 1992 was the precursor to the declaration of Agenda 21, the first international document to introduce ESD. The document recognizes education as an agent for promoting sustainable development. Agenda 21 is an action plan that contains a comprehensive set of principles to assist governments and other institutions in implementing sustainable development policies and programs in their respective countries. Based on Agenda 21, various terms that focus on ESD have been suggested by researchers. Among these suggestions are education for sustainability (Sterling 1996), sustainability education (Wals 2010), sustainable education (Sterling 2001) and education for sustainability (Paulus 1996). Although these terms have minor differences in terms of emphasis, they are interchangeable and reflect the same goals as ESD.

Meanwhile, the SDGs launched by the United Nations are also promoting ESD through a fourth goal dedicated to promoting quality education. According to the United Nations General Assembly (2015: 21), one of the goals was to declare that everyone should acquire the knowledge and skills needed to support sustainable development whereby by 2030, all students must ensure they have the knowledge and skills needed to promote sustainable development, among others, through ESD and sustainable lifestyle, human rights, gender equality, promoting a safe and non-violence culture, global citizenship and appreciation of cultural diversity and cultural contributions to sustainable development.

ESD focuses on new visions and directions for learning and action that help people of all ages to be responsible for enjoying and creating a sustainable future (Haan et al. 2010; Khataybeh et al. 2010). In general, the learning process is found to be more important than the right learning content (Sterling 1996). Sustainable lifestyles cannot be defined in detail because their context depends on time and location. Therefore, a universal curriculum for ESD cannot be established (Wals 2010). Instead, ESD needs to be contextualized and reflect local needs that can be linked to global sustainability issues (Hofman 2015).

### 4.1 Approaches

To date, ESD has emerged as a unifying theme for many types of education and focuses on different aspects of sustainability. In addition, ESD is seen as a catalyst for innovation in education where ESD is increasingly emphasized in almost every country and is seen as a relevant approach to the global problem. As such, various networks and structures have been created at different levels, whether in schools, universities, communities and the private sector to develop ESD.

Mehlmann et al. (2010) state that at the school level in Ukraine, a new integrative and inclusive curriculum was designed for the purpose of ESD to offer additional material for an existing subject. In their teaching, teachers in Ukraine have linked the concept of sustainable development to the field of natural science. However, the social aspect has also attracted the attention of ESD as a sustainable society cannot function without democracy, continuous dialogue, participation and empowerment of society. From a standard school curriculum perspective, social aspects have been revealed through social studies, social psychology, social engineering or philosophy.

In China, cities such as Beijing, Shanghai, Jiangsu, Guangzhou and Mongolia are pioneering the implementation of ESD. The government has also initiated ESD in the education system through the development of ESD curriculum and institutions in teacher education. For prospective teachers attending State University of Education, guidelines on ESD and textbooks as a learning resource have been developed. Meanwhile, the Beijing Academy of Science Education, UNESCO and the Chinese Ministry of Education have developed a "Environment, Population and Education for Sustainable Development" program with the aim of engaging students in activities that reduce social, economic, environmental and cultural problems for sustainable development. In addition, in China, secondary education is considered an important part of ESD practice (UNESCO 2012).

In addition, Manitoba is one of the regions in Canada that urges schools in the region to focus on sustainable development. The Ministry of Education of Manitoba has emphasized ESD as part of its mission statement. In addition, the ministry's main goal is also to ensure that education in Manitoba supports students in terms of experience and learning about a more sustainable lifestyle (UNESCO 2012). Manitoba Education and Training (2000) states that Manitoba has also developed a guidebook for curriculum developers, teachers and administrators focusing on ESD as a means of promoting teaching on sustainable development.

In addition, in the Swedish school system, the concept of sustainable development is not included as a separate subject, but rather as part of the statement for all subjects in the curriculum. Therefore, all teachers are expected to emphasize sustainability issues in the daily teaching process. In addition, sustainable development is considered one of the comprehensive perspectives of education in Swedish schools that need to be outlined (UNESCO 2005). Finland promotes sustainable development by integrating the national curriculum in elementary, general and vocational education at the secondary level. The National Board of Education is also working with schools to improve ESD teaching and learning (UNESCO 2012).

In Jordan and Malta, ESD is integrated into the curriculum of science education and other subjects by introducing the concept of sustainable development through complete pictures and lessons. ESD is also emphasized in the high school curriculum of Italy. However, most education projects on sustainable development are carried out outside of the formal curriculum framework, in collaboration with other environmental NGOs, institutions and organizations (UNESCO 2012). In addition, the Brunei Ministry of Education did not develop an ESD-specific framework and did not establish environmental education as a single subject. However, initiatives have been taken to integrate environmental issues across a range of subjects. In Indonesia, ESD is implemented in collaboration with the Ministry of Environment and the Ministry of Education (UNESCO 2011).

Although ESD has been successfully implemented in many countries, there are some countries that still expect government encouragement and involvement in the implementation. According to UNESCO (2012), countries such as Uganda have stated that there is a need for the government and major ministries to focus on ESD as part of education policy by building on existing school-based ESD approaches. China also hopes that in the future, more attention should be given especially to providing ESD training to teachers and principals in less developed areas. Similar views have also been observed in countries, such as Egypt and the Republic of Korea, where teacher training and guidance are important for promoting teacher and student engagement in ESD (UNESCO 2012).

ESD is interpreted with many interpretations around the world. However, there is a greater recognition of the need for localized interpretations that play an important role in achieving sustainable national status through the process of teaching and learning. In this regard, Malaysia is also not left behind in implementing ESD through various subjects in education. The Ministry of Education Malaysia (MOE) has taken important steps to ensure that education development plans are more practical, realistic and action-oriented, as well as responsive to the needs of the country. The Malaysian education system has been developed in line with national development whereby it is important that ESD is also defined according to the needs of a country. The definition of ESD in the Malaysian context covers aspects of preserving future generations' well-being and what we have today for future generations. To improve the quality of life of the nation, education plays a fundamental role in the development. Therefore, the ministry intends to adopt a holistic approach to human capital development comprising knowledge, skills and ethical values to create progressive thinking and cultural, social and environmental awareness (UNESCO 2011). By 2020, Malaysia aspires to be a developed country that has achieved economic, political, social, spiritual and cultural development. To achieve this, education plays a vital role in the sustainability of the nation.

Various studies have been conducted focusing on different areas to assess the level of integration, implementation and awareness of sustainability issues among students and teachers. Shari and Jaafar (2006) have evaluated the degree of integration and implementation of sustainability issues in the curriculum at architectural schools in Malaysia. The findings show that sustainable design strategies implemented in Malaysian architectural studios focus more on issues related to energy and the environment. However, the study found that there is a lack of attention to social and economic dimensions. However, technology, history and theory and practice and management courses found that technology courses cover more sustainability issues than the other two. Based on an analysis of the integration of sustainability concepts in studio and non-studio teaching, it is recommended that a more balanced approach to sustainability in architectural education should be taken.

In terms of formal education, Ministry of Education Malaysia has played an important role in developing the curriculum for environmental education and introducing a variety of teaching and learning strategies. Curriculum-based environmental education has been introduced in primary and secondary schools since 1998, and the ministry also provides a guidebook on environmental education. Although the ministry has provided the necessary manuals for all schools, its implementation has been limited and unbalanced. This is because the handbook is rarely used by teachers. In addition, some teachers were also unaware of the existence of the handbook. In addition, teachers are of the view that while the handbook suggests and describes activities relevant to environmental education, the details of these suggestions are not sufficient to carry out environmental education programs (Pudin et al. 2004). Said et al. (2007) conducted a study to measure the level of environmental awareness, knowledge and involvement of high school students in sustainable development practices. The results show that students are aware of the increasing environmental problems, but the students' concerns are at a moderate level with regard to environmental issues. Therefore, it is important to look for steps to enhance the understanding and participation of Malaysian students in environmental education and sustainable development.

Karpudewan et al. (2009) have introduced green chemistry experiments integrated with the concept of sustainable development, Sustainable Development Concepts (SDCs) as an alternative approach to teaching existing courses with environmentally responsible modes while taking into account of social and economic aspects. The findings show that green chemistry can enhance students' understanding of traditional environmental concepts, Traditional Environmental Concepts (TECs) and even SDCs. However, from the students' perspective, the understanding of SDCs is much higher than that of TECs. The findings of this study serve as a good resource for student teachers to engage in science education programs and embed scientific concepts based on economic, environmental and social concepts. Furthermore, Minghat and Yasin (2010) have identified aspects of sustainable development in technical and vocational subjects that are expected to guarantee success in achieving sustainable development concepts whereby a sustainable development framework for technical and vocational subjects in secondary schools in Malaysia is proposed to be a guide in integrating sustainable development concepts. This effort is expected to show great potential in the development and conservation of human capital in Malaysia.

Abdullah et al. (2011) state that the introduction of a handbook for teachers to implement environmental education across the curriculum in all subjects shows the importance of environmental literacy to students in achieving environmental sustainability goals. However, the study found that students' level of knowledge about the environment is still poor. Therefore, a study was conducted to determine how well the curriculum in biology, physics and chemistry has integrated environmental knowledge in its core content. The findings show that environmental knowledge is provided only at the surface level in the current science curriculum. Some important aspects of environmental knowledge need to be discussed in depth and applied in science teaching and learning based on consideration of current environmental issues.

Sivapalan (2016) conducted research to reflect, explore and compare the perspectives and beliefs of higher education stakeholders on how sustainable development competencies can be incorporated into Malaysian undergraduate engineering education curriculum. This is because limited research is being conducted to propose holistic guidelines or institutional frameworks for the establishment or evaluation of engineering education for sustainable development at a young age in Malaysia. The guidelines are intended to incorporate holistic sustainable development competencies in the programs of undergraduate engineering and general module learning, and the institution's overall framework is expected to benefit stakeholders, in terms of curriculum and pedagogy, universities and engineering education in Malaysia as a whole.

Peter et al. (2016) have studied the extent to which ESDs are integrated at the community college level in Malaysia. This study found that seven dimensions of sustainability in higher education institutions are discussed, namely curriculum, research and scholarship, operations, faculty and staff recruitment, development and reward, outreach and service, student opportunities, and institutional mission, structure and planning. The findings show that to some extent, the concept of sustainable development can be integrated into three out of the seven dimensions of sustainability, namely curriculum, outreach and services as well as institutional, structural and planning mission.

In addition, Malaysia has also implemented a sustainable school program aimed at promoting awareness of sustainability among students. The program is also known as the Sustainable School Environmental Program and has been in operation since 2005. It is open to primary and secondary students. The results show that interventions on ESD through the program at national level have a more positive impact on the context of participation, responsibility and environmental care indirectly increase ESD awareness. Thus, demonstrating that engagement in these ongoing activities can enhance students' and teachers' awareness of sustainable development and suggesting that the government's efforts to implement the program should be continued throughout Malaysia (Mahat and Idrus 2016). Although the program has been in existence for more than a decade, its curriculum is more environmentally friendly and implemented across curricula covering a wide range of subjects, such as science, geography, civic education and Malay as well as science-based subjects, such as physics, chemistry and biology. Karpudewan et al. (2011b) found that various teaching and learning strategies have been proposed in the syllabus description through environmental education. The aim is to increase the importance of environmental protection among students. However, it is not taught as a specific subject, but these environmental aspects are incorporated into the syllabus at the end of certain chapters.

In addition, Karpudewan et al. (2011a) argue that there is still no specific curriculum developed with a focus on ESD or green chemistry for school students. Therefore, a proactive action is taken by Karpudewan et al. (2011a) by developing a green chemistry curriculum that can support the integration of ESD concept as a course. However, this curriculum is applied to the teaching of chemistry trainees in order to educate and demonstrate more responsible behavior toward environmental sustainability.

## 4.2 Concept

ESD has found its way into the sustainable development movement since the 1990s incorporating concerns about economic and human development for environmental protection (McKeown and Hopkins 2007). ESD needs to provide a scientific understanding of sustainability simultaneously with an understanding of the values, principles and lifestyles that will lead to the transition toward sustainable development (UNESCO 2005). As such, Marks and Eilks (2009) state that science education must move beyond scientific problem solving and application to include socio-scientific decision-making capabilities, thus providing educated citizens who can participate responsibly in the real world. Malaysia has sought to promote awareness of the concept of sustainable development. However, more efforts need to be made in STEM to promote environmentally responsible, economically and socially responsible cultures.

Venkataraman (2009) suggests that ESD throughout the education system are important for the development of people who practice sustainable development as a principle and guide in daily life. However, Burmeister et al. (2012) state that ESD requires a more holistic approach whereby ESD not only requires the use of new questions about the concept of sustainability or the network of science and society as the content and/or context of science teaching, but more detail to enhance understanding. In addition, ESD requires a more comprehensive approach to relate social issues and multidimensional management. Diversity in this dimension should include an understanding of the background of the issues presented where it can also come from chemistry, and that perspective should also be introduced in compulsory science education in schools. At the same time, chemistry education based on ESD principles at all levels must also be oriented toward addressing the ecological, economic and social impacts as a whole, while focusing on real change in society at the local, regional and global levels (Haan 2006; Wheeler 2000).

Considering the concept of chemistry education in schools, the study of chemistry theory and facts alone cannot enhance students' ability to understand the concept of sustainable development. The use of more community-oriented approaches from multiple dimensions through chemistry education can provide the necessary encouragement to achieve educational goals for development (Ware 2001). With these improvements, education has become a leading field for learning how chemicals are embedded in people's daily lives including their environmental, economic and social impact (Hofstein et al. 2011). In addition, learning about how chemistry development is linked to the environmental, economic and social impacts and understanding that arises from these issues is more important. Thus, chemistry education has been shown to have important potential for improving the level of general education skills among students from a more participatory learning perspective. This is because recent societal developments are directly linked to chemistry and technology which are then managed through a multidimensional approach (Burmeister et al. 2012).

Yencken et al. (2000) suggest that daily experience and knowledge as well as three dimensions of sustainability, namely economic, environmental and social can be reflected in the curriculum by integrating sustainable development concepts. Among the 20 proposed sustainable development concepts are carrying capacity, steady-state economy, ecological space, sustainable development, ecological footprint, natural resource accounting (natural resource accounting), eco-efficiency, life cycle analysis, sustainable consumption, 5 Rs. (the 5 Rs.), local global links, interdependence, biodiversity, interspecies equity, intra-generational equity, intergenerational equity, human rights, basic human needs, media literacy and democracy. All these proposed sustainable development concepts are expected to assist either researchers or teachers in integrating sustainable development concepts.

Karpudewan et al. (2011b) have adapted the integration approach suggested by Yencken et al. (2000) in the method of teaching chemistry offered to pre-service chemistry teachers. This study outlines new teaching strategies that contain scientific, environmental and social science concepts that will be a good resource for trainee teachers involved in science education programs especially green chemistry education. The results show that the integration of sustainable development concepts in green chemistry subjects can influence the value change among trainee teachers. It is important that these future teachers understand the right values because it is important for teachers to have a tendency to articulate values that are important to students through knowledge of pedagogical content (Veugelers 2000). In addition, Heaton et al. (2006) also suggest that the integration of green chemistry principles can assist in curriculum development and pedagogy that will lead to the development of values, knowledge and skills that can contribute to sustainable development.

According to Millar (2008), science education in schools aims to provide students with a multidimensional understanding of science in which students can apply the knowledge gained in personal, daily and community life. At the same time, Holbrook and Rannikmae (2009) state that one of the important goals of scientific literacy is to provide students with the ability to make responsible decisions based on their knowledge and ability to interpret, understand and apply relevant scientific concepts and ideas. Global and local perspectives state that it is important that education should focus on what will be truly useful and meaningful to every individual in the community in the future. Landorf et al. (2008) define education for sustainable human development as an educational practice that results in improvements in human well-being, both for the present and for future generations. However, Wals (2007) argues that ESD also needs to establish a close link between education and real-life experience by focusing on sustainability issues encountered in a society itself. In general, ESD aims to help students in developing the attitude, skills and knowledge needed to act on informed decisions for themselves, society and future generations (McKeown et al. 2002).

According to Venkataraman (2009), the inclusion of ESD throughout the education system is important for the development of people who practice sustainable development as a principle and guide in their daily lives. Therefore, McKeown et al. (2002) argue that if the basic level of education can be improved, the second priority of ESD is to reconsider elementary and secondary education to address issues of sustainability. In contrast to environmental education as a separate entity in the curriculum, McKeown et al. (2002) state that ESD skills, knowledge and values are reflected in the learning system. Similarly, in defining ESD, UNESCO (2005) has stated that the skills, values and practices of sustainable development need to be integrated in all aspects of education and learning. As such, the orientation of the chemistry curriculum shows a positive tendency to adopt a more sustainable lifestyle and this is a good starting point for educators to develop the curriculum. In this regard, ESD-related frameworks that can help to integrate the process need to be taken seriously.

A conceptual framework has been developed based on previous research and UNESCO documents related to the integration of sustainable development concepts toward achieving ESD. Models of knowledge, attitudes and behaviors (Ramsey and Rickson 1976) and Hines's model of Responsibility to the Environment (Hines et al. 1987) are used. The pro-environmental model of behavior is the earliest and simplest model proposed to measure and to explain environmental awareness and concern (Ramsey and Rickson 1976). This model has been widely accepted because it associates knowledge with attitudes and attitudes toward behavior. This model is then known as the knowledge, attitude and practice (KAP) model (Mahmud and Siarap 2013). Vandamme (2009) states that the KAP-based survey has received criticism in terms of reliability, validity and measurement related to the intensity of opinions or attitudes. However, KAP analysis is well accepted as a conceptual framework for assessing people's understanding, awareness, readiness and participation in a particular issue (Launiala 2009). Kollmuss and Agyeman (2002) also agree that most non-governmental organizations (NGOs) still have a simple assumption that knowledge depth will lead to behavioral excellence. In general, studies related to KAP are found to be useful for three general purposes, as a diagnostic tool to illustrate current knowledge, attitudes and practices of society to communicate current situation and plan appropriate interventions and as a tool for assessing the effectiveness of specific interventions or programs (Vandamme 2009).

Recently, various KAP-based studies have been widely used to determine the level of human behavior in various fields. These include food security, education (Idros 2006), management (Laor et al. 2018) and public health

(Launiala 2009). However, in the field of ESD, the KAP survey has been widely used to assess the level of knowledge, attitudes and practices or behavior of academic staff and students on issues related to awareness of sustainable development concepts (Azapagic et al. 2005), engineering education (Sivapalan 2016), campus sustainability (Choy et al. 2017; Wan Nur'ashiqin et al. 2011), the field of architecture (Emanuel and Adams 2011) and environmental sustainability (Idros 2006). Flamm (2006) states that the KAP model is also known as the knowledge, attitude and behavior (KAB) model where knowledge is given priority as the basis for determining attitudes and behaviors. Based on the KAP model, an increase in an individual's knowledge leads to a change in attitude. Changes in attitude will lead to changes in practice or behavior. Knowledge can be enhanced through exposure to new information through lectures, classes, media, lectures and other scientific activities (Isa 2016). Iyer (2018) argues that knowledge of concepts is important for a person to be positive. A positive attitude can lead to positive behavioral changes by organizing activities that can motivate peers and others. These behavioral changes are expected to be lifelong. Figure 1 shows the earliest model of pro-environmental behavior.

In 1986, Hines, Hungerford and Tomera introduced the Hines Model of Responsible Environmental Behavior. This model is proposed based on the Theory of Planned Behavior proposed by Ajzen and Fishbein (1980) as shown in Fig. 2. Hines model was introduced after conducting a meta-analysis on environmentally responsible behaviors and studies used to describe processes involved in fostering positive environmental behaviors through environmental education (Kollmuss and Agyeman 2002). The model identifies six variables, namely

(i) Knowledge of issues: One needs to know about environmental issues and their causes.
(ii) Knowledge of action strategies: One needs to know how to act to reduce the impact of environmental problems.
(iii) Locus of control: This represents the perception of the individual whether the individual has the ability to bring about change through his or her own behavior. Someone with a strong locus of internal control believes that their actions can lead to change. However, people with the locus of external control find their actions insignificant and feel that change can only be taken by more powerful people.
(iv) Attitudes: People with strong pro-environmental attitudes are found to be more likely to engage in pro-environmental behaviors, but the relationship between attitudes and actions has been shown to be weak.
(v) Verbal commitment: Commitment willingness to take action also provides some indication of a person's willingness to engage in pro-environmental behavior.
(vi) An individual's sense of responsibility: A person with a higher personal responsibility is more likely to engage in environmentally responsible behavior.

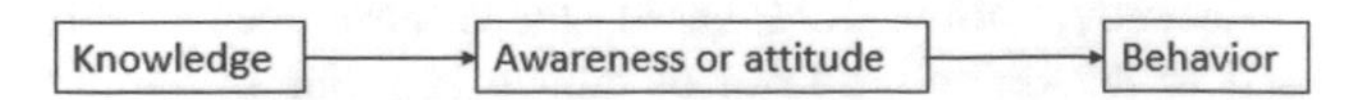


**Fig. 1** Earliest model of pro-environmental behavior (Ramsey and Rickson 1976)

Based on this model, individuals who wish to take action are more likely to engage in environmental-related actions than individuals who do not have such desires. Before a person wants to take action to address a specific environmental issue, the individual must be aware of the existence of the problem. Therefore, knowledge of an issue is seen as a prerequisite for action. One must have knowledge of available behaviors in order to be more effective in certain situations. According to Davey (2012) and Kollmuss and Agyeman (2002), there are several situations or situations that influence environmentally responsible behavior. Situational factors in this model refer to constraints, social pressures and opportunities to choose different actions that may hinder or enhance one's desire to act. Therefore, this model is expected to predict one's behavior to be more environmentally responsible.

Based on the KAP model and Hines model, a conceptual framework in which the main variables of the concept includes three categories of variables and one category of variables that hinder the path to action. In this conceptual framework, the three main domains that refer to the KAP model are adapted to assess the level of awareness and knowledge, attitudes and behaviors as each aspect of awareness will begin with knowledge and subsequently result in changes in attitude and behavior as well as practice. In a study conducted by Sahin et al. (2012) at the Middle East Technical University on the attitude of university students toward the concept of sustainability, it was found that knowledge of environmental issues affects individual attitudes toward the environment. This attitude leads to behaviors that may encourage individuals to deepen their knowledge of campus sustainability. At the College of Alabama and Hawaii, Emanuel and Adams (2011) conducted a study to determine students' perceptions of campus sustainability. The study found that students' understanding, and perceptions of sustainability can provide insights into ways to practice sustainability in their students' daily lives whereby deep understanding can only be obtained through knowledge. Furthermore, Chen et al. (2011) argue that in achieving the sustainability objectives, changes in the

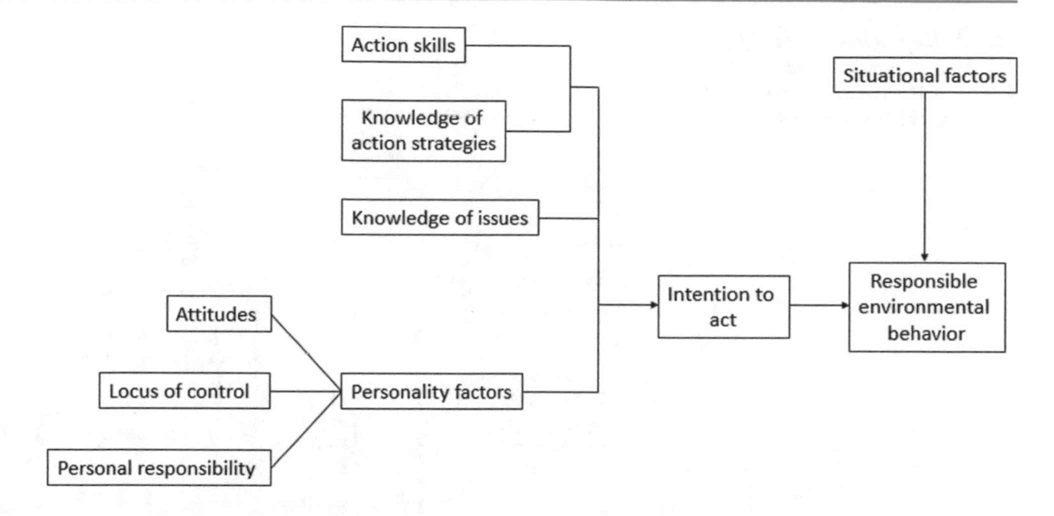


**Fig. 2** Hines model of responsible environmental behavior

attitudes and behaviors of individuals at both faculty, staff and students are important. Arbuthnott (2009) has stated that attitudes that focus on sustainability actions where ESD programs include strategies that target personal or professional characteristics in the intention can assist in transforming the intention into action. Therefore, the alignment of ESD efforts toward decision-making in terms of institutional infrastructure, regulation and incentives will have a profound impact on more sustainable behavior.

The integration of sustainable development concepts is generally influenced by several barriers that can be generalized into six key areas, namely education policies, curriculum orientation, pedagogical approaches, resources, stakeholders and awareness, that may influence the process of integration of sustainable development concepts (Kanapathy 2018, 2019a, b). Figure 3 shows the conceptual framework is based on KAP model and Hines model that illustrates levels of knowledge, attitudes and behaviors toward sustainable development dimensions, namely economy, social and environment in the context of chemistry curriculum are dependent on education policies, curriculum orientation, pedagogical approaches, resources, stakeholders engagement and awareness.

a. Education policy

In general, educational policy pushes toward goals and procedures for achieving goals (Wright 2006). To fully integrate the concept of sustainable development into the curriculum, there is a need to integrate ESD principles in national education policy. In addition, shifting current educational goals by incorporating ESD principles or shifting national education policy aspects can accelerate the process of integrating sustainable development concepts (UNESCO 2015). In recent years, the existence of policies to support sustainability initiatives has been relatively rare; despite its policies, the enforcement of these policies has been found to be less or less effective in guiding day-to-day activities (Wright 2006).

Education policies, strategies and documentary guidelines can guide and define educational priorities and orientations. In order to integrate sustainable development concepts through curriculum, policies, strategies and specific policy guidance documents need to be reviewed. These include policies related to curriculum organization, teaching and learning strategies, assessment, classroom learning materials, school-community relationships and professional development among teachers. All these policies are designed to indirectly influence the learning system in terms of the knowledge, values, attitudes and skills that teachers convey to students (UNESCO 2015).

b. Curriculum orientation

The curriculum available does not include all relevant and essential elements of sustainability (UNESCO 2015). In addition, the complexity of the scientific concepts involved can also limit the understanding of the principles of sustainability among students. Therefore, the development of appropriate curriculum equipping students with knowledge of the concept of sustainable development can recognize that human behavior results in normative environmental complexity. In addition, there are studies show that students with the courage, the knowledge to explore and the desire to act are more likely to contribute toward a more sustainable future (Strauss 1996) where understanding the issue of sustainable development is essential to finding solutions (Hayles and Holdsworth 2005).

c. Pedagogical approach

Most literatures focus on pedagogical approaches to achieving ESD (Cotton et al. 2009). Traditional teaching methods have been identified as less suitable approaches for

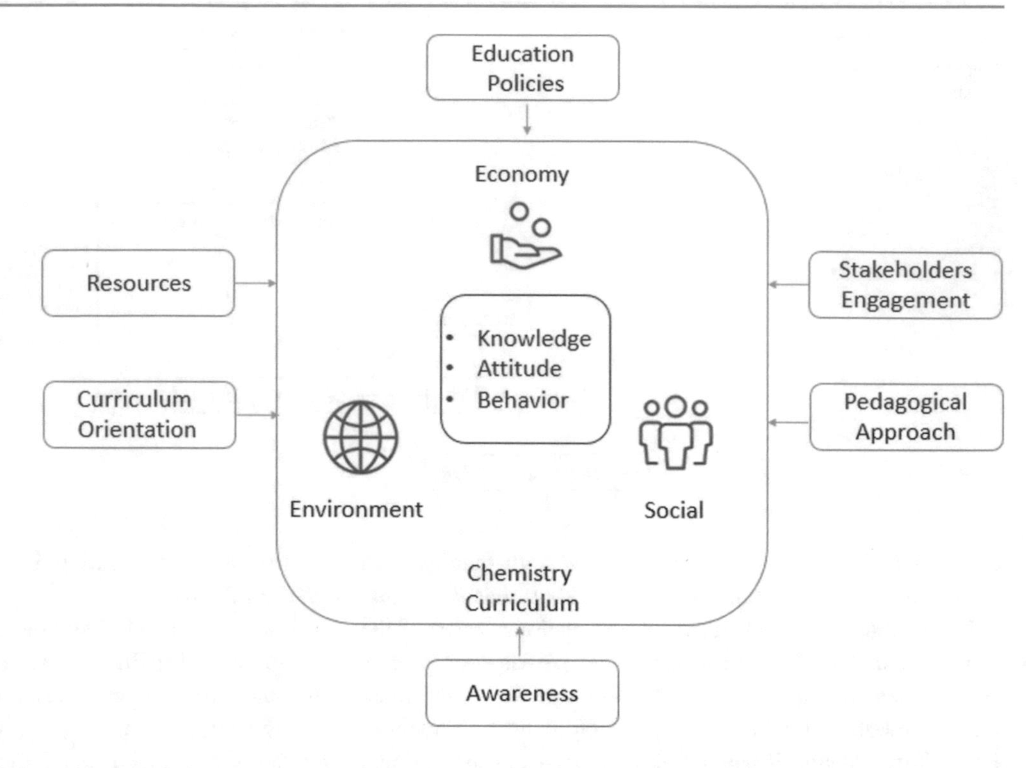


**Fig. 3** Conceptual framework adapted from KAB model and Hines model (Ramsey and Rickson 1976; Hines et al. 1987)

teaching sustainable development concepts because traditional teaching is found to be less helpful to students in understanding the real issues of everyday life based on the knowledge gained (Papadimitriou 2004). ESD's focus is on the "what" and "how" of real-world problems where a more innovative approach to pedagogy is needed (Ryan and Cotton 2013). As such, there is a need to encourage teaching approaches that design, make decisions and focus on problem solving techniques and solution-seeking (Hayles and Holdsworth 2005).

The findings of previous studies have emphasized that the integration of sustainable development concepts requires a paradigm shift in the pedagogical approach which means the transition from transmissive learning to discovery learning; teacher-centered approach to student-centered approach; learning that focuses the theories on more practical learning which combines theories with practices (Sterling 2004). In addition, most researchers have also focused on transformative sustainable learning (Mintz et al. 2013; Sipos et al. 2008; Wals 2010). Transformative sustainability learning is a combination of transformative learning with ESD that focuses on cognitive, psychomotor and affective domains in which integration of these three domains can influence student behavior (Sipos et al. 2008).

In addition, by engaging students through pedagogical approaches based on student centralization, such as case study, learning by doing, service learning, problem-based learning, experiential learning, active learning and participation, self-directed inquiry, engagement with real-life problems and related issues, collaborative learning in the community, are suggested to enhance students' real-world learning opportunities (Figueiro and Raufflet 2015; Moore 2005a; Thomas 2004). However, among the proposed approaches, participatory, active and collaborative learning have been identified to provide the most meaningful learning experience for students. This is because the learning techniques encourage students to choose STEM fields by integrating socio-scientific issues in science education, students' motivation and interest in learning can be enhanced (Sadler and Dawson 2012; Tal et al. 2011; Tal and Kedmi 2006; Terenzini et al. 2001).

d. Resources

Some studies suggest that resources, such as time, finance, expertise and training are factors that hinder the implementation of sustainable development concepts (Chakraborty et al. 2016; Filho 2000; Jorge et al. 2015; Velazquez et al. 2005; Wilson 2012). In the eyes of teachers, time is a limited resource because any improvement or change in syllabus will affect a teacher's work attitude and orientation (Hargreaves 1994). Most teachers around the world want to develop interdisciplinary teaching and learning, but these teachers feel that the need to follow and complete the existing syllabus that causes time constraints (Borg et al. 2012). This opinion is in line with the findings of the study by Burmeister et al. (2012) that state a cohesive curriculum has a timely impact on innovation and discussions with colleagues. In addition, lack of time also influences formal planning, evaluation and reporting processes to enhance sustainability issues in the curriculum that are typically found (Velazquez et al. 2005).

Financial resources are of the main concerns of teachers as they are not only increasing the risk of achieving all their goals, but achievements of future sustainability-related initiatives (Velazquez et al. 2005). In addition, at the university level, finance is distributed on the basis of priorities where integration of the sustainable development concepts through the subject is not viewed as an important priority causing the financial allocation of sustainability-based activities to be less focused (Filho 2000). Cebrian et al. (2015) suggest that the existence of specific financing schemes can encourage the integration of sustainable development concepts into the curriculum among teachers.

In addition, initiatives for the integration of sustainable development concepts are also limited by the lack of experts in the field of sustainability. This has caused the majority of academics to misunderstand the concept of sustainable development (Cebrian et al. 2015). In fact, only a few teachers are exposed and taught how to integrate the concept of sustainable development in order to gain knowledge of the field (Velazquez et al. 2005). Therefore, ESD professional development and training programs for academics are important to provide time and opportunity to gain insights and knowledge on sustainability, rethink the relevance of existing teaching practices and acquire the skills needed to teach on sustainability issues (Thomas 2004).

e. Stakeholders engagement

Steps toward integrating sustainable development concepts into the curriculum require cooperation from various stakeholders. Implementing ESD in the curriculum is a complex system that involves stakeholders from various levels, including management, academics, students and parents. Therefore, cooperation between all stakeholders needs to be strengthened in order to successfully implementing ESD (Blum et al. 2017).

Most studies related to the concept of sustainable development focus on students' awareness, knowledge, attitudes and behaviors (Emanuel and Adams 2011), senior management (Lozano 2006) and teachers (Cebrian et al. 2015). In addition, there are also several studies that look into the perspectives of parents and alumni students on the concept of sustainable development. Parents of students are suggested as one of the key external stakeholders that need to be involved in studies related to the integration of sustainable development concepts at school level (Disterheft et al. 2012).

To encourage the integration of sustainable development concepts, an integrative functioning organizational structure is proposed (Viebahn 2002). This is because any decision on policy, curriculum or resource-related changes can be made more quickly by management (Karabell 1998) where organizations play a role in explaining ESD's importance and ESD implementation requirements verbally and through action at all levels. In addition, appropriate strategies to engage academics in ESD are positively needed to prevent ESD being considered as mandatory agendas (Hegarty 2008).

Besides that, initiatives to engage students in achieving ESD goals have been focused either by integrating sustainable development concepts into public institution activities (Biedenweg et al. 2013) or through teaching and learning (Adomßent et al. 2014; Labodova et al. 2014). For example, sustainability education has taken steps to understand and achieve the concept of sustainable development by engaging students in activities, such as campus conservation initiatives, field trips, environmental courses and workshops (Matsuura 2004). However, in order to implement all these initiatives effectively, various challenges need to be met and require input from various stakeholders, including students, academics, management, non-governmental organizations and parents (Barth and Rieckmann 2012).

f. Awareness

The concept of sustainable development is a proactive and practical futuristic paradigm considering environmental, economic and social balance in development planning and improving quality of life as well as addressing sustainability issue (McKeown et al. 2002; Sterling 2004). In order to achieve sustainable development and to build a more sustainable community with the principles of sustainability, the balance between economic, social and environmental needs should be given attention (Zamhari and Perumal 2016). However, studies show that the understanding of the precise definition of sustainable development is still poorly understood and that awareness of the concept of sustainable development is still at an unsatisfactory level (Mahat et al. 2013a).

For example, in Manitoba, Canada, the government has taken initiatives to assist teachers in ESD or Educating for Sustainability (EfS) in the areas of knowledge, skills and values that contribute to sustainable development (McDonald 2006). In addition, a study conducted in Australian primary and secondary schools on teachers found that only three teachers are familiar with ESD. Although in general, teachers acknowledge the importance of education in achieving sustainable development, it is the lack of understanding that causes them to hesitate to do so (Taylor et al. 2003). Mahat et al. (2013b) argue that awareness can be enhanced if the knowledge and understanding of an individual can be applied in positive behaviors especially through more sustainable practices in daily life. Therefore, in order to create sustainability for future generations, individuals need to be exposed to appropriate knowledge, skills and values and one of the agents that can successfully execute them is education (Burmeister and Eilks 2013).

The 2030 agenda is a holistic plan of action comprising social, environmental and economic dimensions. To understand the concept of sustainability, one must take into account these three main dimensions, which are the "Three Stages of Sustainability." These three dimensions are interrelated and, when combined and applied in real-world situations, can provide a solid foundation for a more sustainable world in which universal society can benefit (Vasiliki and Maria 2015). To achieve environmental sustainability, there is a need to maintain the functions and utilities of the natural environment for a long time. To support this, care in terms of natural environment and at the same time positive growth rates are encouraged. In addition, any actions that interfere with environmental balance should be avoided and in the event of any unforeseen occurrence should be promptly resolved taking into account the environmental impact. Environmental sustainability covers a wide range of issues ranging from pollution to natural resource management (Vasiliki and Maria 2015). Therefore, by protecting the planet from degradation, including through sustainable consumption and production, management of natural resources and taking immediate action against climate change, action plans for the planet can be achieved (United Nations 2015). Sustainability of the economy is the ability of the economy to sustain a fixed level of economic growth. Economic sustainability refers to any decision taken with the utmost caution after thoroughly reviewing other aspects related to sustainability. Economic sustainability also includes a wide range of things, ranging from "smart growth" to subsidies or tax exemptions for green development. In addition, economic sustainability also emphasizes lowering spending that is not important (Vasiliki and Maria 2015). Thus, the action plan for the economy is intended to ensure that the community can enjoy a prosperous and fulfilling life and to ensure that economic, social and technological progress is in keeping with the environment (United Nations 2015). Social sustainability depends on any decision or project that promotes social improvement. In general, the social aspect of sustainability supports the concept of intergenerational justice. This means that future generations have the right to enjoy the same quality of life or higher than the current generation. This concept also covers many other social-related issues. However, the social sustainability dimension shows a similar importance to the other two sustainability principles. If this dimension is not taken seriously, it can lead to the collapse of the entire sustainability process while also impacting the community itself (Vasiliki and Maria 2015).

As such, the relationship and nature of the integration of the 17 SDGs play an important role in ensuring awareness and achievement of the new 2030 Agenda (United Nations 2015). Students' understanding and awareness are essential to successful integration of sustainable development concepts into the chemistry curriculum. The three aspects identified to accelerate understanding of environmental, economic and social sustainability through the chemistry curriculum among students are

(i) Knowledge: Knowledge refers to the basic understanding and search for solutions in real life as well as developing social and environmental responsibilities and assessing values that are conducive to sustainable development.
(ii) Attitude: Attitude is defined as the permanent positive or negative feeling about a person, object or issue (Bell 1998). Attitude refers to self-reflection and appreciation of the importance of social, environmental and economic aspects.
(iii) Behavior: Behavior represents the intention to protect the environment and resources that consider the needs of future generations and at the same time meet current needs.

Various studies involving different fields have analyzed the importance of knowledge and the effects of lack of knowledge in decision-making. Bell (1998) states that the attitude and behavior of an individual are influenced by the level of knowledge. According to Laroche et al. (2001), changes in attitudes and behaviors among students are personal and are usually a slow process. In addition, students' daily lives are found to be closely related to their social environment, personal values, attitudes, interests and motivations. In general, the relationship between attitude and sustainable behavior is complex.

In addition, some studies also suggest that there is a correlation between attitude and behavior, but there are also studies that do not find any relationship between the two variables (Dwyer et al. 1993). Behavior is a case of situations in which it differs from the social environment (Yin et al. 2002). The attitude and behavior of an individual are a state of constant change caused by the process of learning, prioritization and perception. Therefore, it is important to introduce sustainability issues in long-term teaching in all school subjects (Juntunen and Aksela 2014).

More recently, various fields of academic interest have begun to focus on sustainability issues where emphasis is placed on teaching and learning approaches. Many experienced and new teachers are working hard to implement ideas of sustainable development through curriculum due to various challenges. To overcome these challenges, a general framework as proposed is intended to be applied to all subjects related to science, technology, engineering and mathematics that integrate sustainable development concepts into the teaching and learning processes. However, associating the dimensions of sustainable development with the chemistry curriculum is an important issue whereby this shows that the three dimensions are of equal importance. Therefore, having a deep understanding of these three dimensions through chemistry education is expected to lead to significant changes in knowledge, attitudes and behaviors that will be useful in developing an individual's ethics and actions toward sustainable development. Although these three dimensions are important, the three dimensions of sustainable development concepts are not required to be integrated into each chemistry teaching and learning session but teachers are encouraged to have a deep understanding of the relationships between these three dimensions for each topic in the chemistry subject to facilitate the teaching and learning processes.

## 5 Case Study: Chemistry Curriculum in a Pre-university Program

The purpose of science in the context of sustainability is to understand and explain the dynamics necessary to prevent humans as individuals and collectively, physically, socially, economically, culturally and psychologically from destroying the environment (Anon 2018). Although there is a general desire to integrate ESD into science education, studies have shown that learning about sustainability issues is relatively low in secondary education, including chemistry education (Burmeister et al. 2012). Therefore, it is suggested that the national curriculum and publications in science education should strive to strengthen ESD in science education (Osborne and Dillon 2008; Rocard et al. 2007).

The subject of chemical sciences is defined as the foundation of modern life and society (Bradley 2005). Therefore, chemistry education has a special responsibility to contribute to ESD through formal and informal education (Burmeister et al. 2012). Existing chemistry teaching materials focus primarily on the technological or environmental aspects, without reflecting the full ecological, economic and social impact. Additionally, there are discussions on whether to add environmental issues or the basics of chemical technology related to sustainability in chemistry content or contexts. To address this issue, a chemistry curriculum that includes the practice of the Green Chemistry concept, socio-scientific chemistry education and the integration of sustainable development concepts into chemistry education strategies have been proposed (Burmeister et al. 2012).

According to Karpudewan et al. (2012), general knowledge and domain-specific knowledge and skills need to be developed to enable an individual to evaluate new chemical-based products and technologies in their own lives and in the society in which they live and work and respond accordingly. Students also need to develop the same skills regardless of whether they are going to start a career in science and technology. The current generation is the future leaders and needs the ability to move forward to engage in social discussions and make decisions on sustainable development issues. To achieve the ESD goals for all students in chemistry, a variety of different suggestions for integrating sustainability concepts in secondary chemistry teaching have been suggested. This proposal covers changes in content (Eissen 2012) to the overall curriculum change with emphasis on chemistry learning (Burmeister and Eilks 2012).

In addition, chemistry education can also help students to actively understand and participate in social discourse and to make decisions on socio-scientific issues related to chemistry and a variety of other applications (Hofstein et al. 2011). In addition, the preparation of a chemistry syllabus toward developing competencies for students' understanding and allowing them to engage in social discussions on chemistry applications and technologies can make chemistry education more relevant to ESD. To achieve this goal, students must have strong chemistry knowledge in the context of socio-scientific or sustainability issues. The mere attitude of students to acquire knowledge based on chemistry content alone is not sufficient to achieve ESD through the subject of chemistry (Sjostrom and Stenborg 2014).

In this case study, the pre-university chemistry curriculum at the PERMATApintar™ National Gifted Center, Malaysia, is selected because the students are given exposure to view the knowledge links obtained from various subjects with real-life situations (Yassin et al. 2012). In addition, the center emphasizes not only the academic excellence of the students, but also the holistic development of the students as outlined by the National Education Blueprint (Bakar 2017). In addition, Malaysia's recognition of sustainable development through the 11th Malaysia Plan reflects the importance given to the country on sustainable development. The concept of sustainable development highlighted in the context of chemistry above is a guide for chemistry educators who want to collaborate on embedding the concept of sustainable development as part of student life and this initiative is a proof that chemistry education has the potential to contribute to sustainable development.

The center emphasizes student holistic development which includes physical, emotional, spiritual, intellectual and social aspects. Students participating in the program are offered advanced courses in biology, chemistry, physics, mathematics and statistics, which resemble first-year courses at the university level. In addition, students can also conduct research in the field of interest. One of the main goals of the program is to produce the younger generation in line with Malaysia's need for an industrialized nation by 2020 (Pusat

PERMATApintar™ Negara 2018). Therefore, the findings from this case study are expected to yield results that will benefit other pre-university students.

## 6 Enculturing Sustainable Development Concept Through Chemistry Curriculum

The transition toward the integration of sustainable development concepts through chemistry subjects is an important and serious consideration for a pre-university program at the PERMATApintar™ National Gifted Center. However, the integration of sustainable development concepts with chemistry subject cannot be accomplished by simply incorporating them into the syllabus of chemistry subject. In light of this, a framework for integrating sustainable development concepts through the subject of chemistry (Fig. 4) is developed based on the concept as shown in Fig. 3, whereby this case study suggests that there are six aspects that are closely related to each other that need to be focused on to effectively integrate sustainable development concepts through the subject of chemistry, namely (1) education policy; (2) awareness; (3) resources; (4) curriculum orientation; (5) pedagogical approaches; and (6) stakeholder engagement. This framework clearly describes the things that need to be considered at every level before beginning to integrate the concept of sustainable development through the subject of chemistry.

Generally, the integration of sustainable development concepts through chemistry subject depends on education policy at the government and institutional levels whereby the policy is to pave the way toward in achieving the goal of integrating sustainable development concepts through chemistry subject. Without a policy that emphasizes on the concept of sustainable development, any plan to integrate the concept of sustainable development through the subject of chemistry cannot be effectively implemented.

The integration of the concept of sustainable development through the subject of chemistry begins with the awareness of stakeholders, whether external or internal stakeholders within an institution. With awareness, a basic understanding of the concept of sustainable development and an appreciation of the benefits of integrating sustainable development concepts through the subject of chemistry can be enhanced among teachers. However, this can only be done with the help of training resources. In addition, in order to gain an appreciation of the benefits of integrating sustainable development concepts through chemistry subject, pedagogical approach and curriculum orientation need to be emphasized. Throughout this process, aspects of pedagogy and curriculum also need to be focused on deepening the knowledge of how and where to integrate sustainable development concepts through the subject of chemistry. In the meantime, in order to create an ongoing appreciation for the integration of sustainable development concepts in the teaching of chemistry subject and their involvement in teaching the concept of sustainable development in the classroom, teachers need to be provided with resources and time. The six aspects outlined in this framework are important to consider when integrating sustainable development concepts through chemistry subject at school or institution level across the country. This indirectly encourages and engages teachers in integrating sustainable development concepts through the subject of chemistry.

Overall, the implementation of this framework involves the involvement of internal and external stakeholders within an institution. The external stakeholders identified in this framework are the government, the Ministry of Education and non-governmental organizations. For the integration of sustainable development concepts through chemistry subject, the government is directly involved in the formulation of ESD-related policies, while non-governmental organizations are directly involved in assisting the government in implementing ESD-related programs at the institution level. At the institutional level, the involvement of key stakeholders, including administrators, teachers, parents and students, should be emphasized in order to successfully integrate sustainable development concepts through the subject of chemistry. With the help of this framework, specific aspects that need attention can be researched and used as a guide to the process of integrating sustainable development concepts through chemistry subject at schools or institutions nationwide and this indirectly encourages and engages teachers in conducting sustainable chemistry education.

## 7 Conclusions

ESD is a learning process that aims to equip students, teachers and the school system with the new knowledge and thinking needed to achieve economic well-being and act as responsible citizens and restore the environment in which living organisms depend on The Cloud Institute for Sustainability Education (2019). However, the diversity of issues within the ESD invites the involvement of various stakeholders in the implementation of ESD, namely the government, non-governmental organizations and the media. The roles and responsibilities of the various parties vary according to needs; however, the cooperation of all these parties is necessary to leverage the synergy.

Recent developments in Malaysia show a deep interest by the Malaysian government in STEM education. In Malaysian Education Blueprint 2013–2025, Ministry of Education

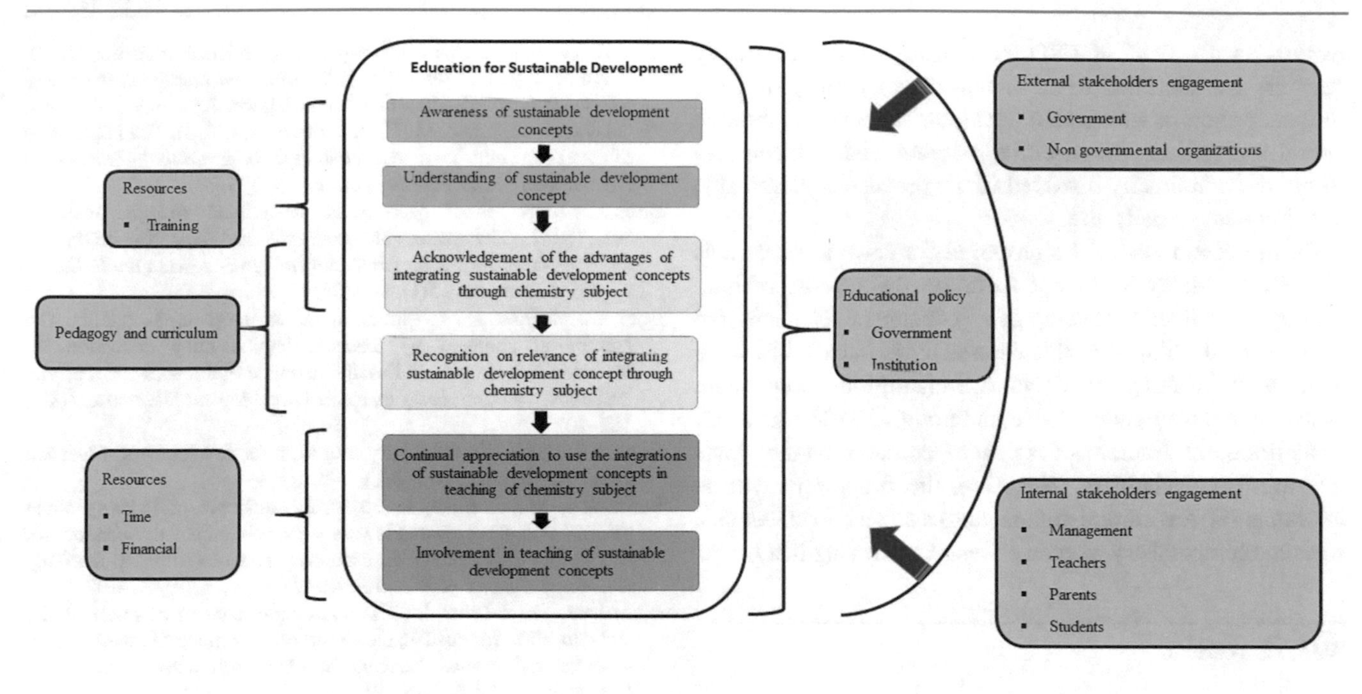


**Fig. 4** Framework for integration of sustainable development concepts through chemistry subject

Malaysia (2013) stated that the government intends to ensure that students are equipped with the skills needed to meet the changing world with the application of Science, Technology, Engineering and Mathematics (STEM). In addition, ESD was introduced to recognize the need to promote sustainable development through integration in the curriculum for the benefit of future generations. However, in Malaysia, the emphasis on integrating sustainable development concepts in STEM education, especially focusing on chemistry, is lacking.

This issue has raised several questions that need to be addressed in relation to the integration of the concept of sustainable development through the subject of chemistry. This case study addresses the integration of sustainable development concepts in the Malaysian education system, particularly in the field of STEM as well as the chemistry subject of pre-university program at the PERMATApintar™ National Gifted Center. Furthermore, these issues also raise questions about the levels of knowledge, attitudes and behaviors of sustainable development concepts in chemistry subject, factors that influence the integration of sustainable development concepts in chemistry subject as well as the planning needed to integrate sustainable development concepts into the subject matter of the pre-university curriculum at the PERMATApintar™ National Gifted Center.

A framework for integrating sustainable development concepts through the subject matter of the pre-university program at the PERMATApintar™ National Gifted Center is developed. The framework suggests that to effectively integrate the concept of sustainable development through the subject of chemistry, there are six main areas that need attention, namely education policy, awareness, resources, curriculum orientation, pedagogical approach and stakeholder's engagement. Two aspects of policy need to be considered are government and institutional policies. Meanwhile, the aspect of awareness is concerned with understanding the concept and relevance of the concept of sustainable development through the subject of chemistry. In addition, the types of training, finance and time have been identified as aspects of resources that need to be considered in integrating sustainable development concepts through chemistry subject. The integration of sustainable development concepts in syllabus and the contextualization of chemistry content as aspects that should be considered in the curriculum. In the curriculum, learning outcomes need to be revised to facilitate the process of integrating sustainable development concepts through chemistry subjects. Pedagogical approach needs to be considered and this includes pedagogical types, learning approaches and pedagogical techniques. Subsequently, stakeholders include government, experts in the field of ESD, administrators and chemistry teachers are identified to be directly involved in promoting the integration of sustainable development concepts through chemistry subject. Meanwhile, parents and students are found to be indirectly involved in the practice of sustainable development in daily life.

In the past, most of the proposed frameworks or models of ESD in Malaysia do not focus on the process of integrating sustainable development concepts. As such, the framework developed in this case study is intended to serve as a guide for the government and institutions to enculture sustainable development concepts through STEM education. In addition, the framework is developed based on the inputs by various stakeholders. Therefore, the framework will be instrumental for administrators and teachers to develop a sustainable chemistry teaching toward achieving ESD.

## References


Abdullah, S. I. S. S., Halim, L., & Shahali, E. H. M. (2011). Integration of environmental knowledge across biology, physics and chemistry subject at secondary school level in Malaysia. *Procedia-Social and Behavioral Sciences, 15*(1), 1024–1028.

Abidin, N. Z. (2010). Investigating the awareness and application of sustainable construction concept by Malaysian developers. *Habitat International, 34*(1), 421–426.

Anon. (2018). Ministry of Education-empowering youth's trough science and technology. https://www.guyanatimesinternational.com/ministry-of-education-empowering-youths-trough-science-technology/ viewed 2 Mac 2018.

Adomßent, M., Fischer, D., Godemann, J., Herzig, C., Otte, I., Rieckmann, M., et al. (2014). Emerging areas in research on higher education for sustainable development–management education, sustainable consumption and perspectives from Central and Eastern Europe. *Journal of Cleaner Production, 62*(1), 1–7.

Agenda 21. (1992). United Nation Sustainable Development. United Nations Conference on Environment and Development, pp. 320–328.

Ajzen, I., & Fishbein, M. (1980). *Understanding attitudes and predicting social behaviour*. New Jersey: Englewood Cliffs.

Arbuthnott, K. D. (2009). Education for sustainable development beyond attitude change. *International Journal of Sustainability in Higher Education, 10*(2), 152–163.

Azapagic, A., Perdan, S., & Shallcross, D. (2005). How much do engineering students know about sustainable development? The findings of an international survey and possible implications for the engineering curriculum. *European Journal of Engineering Education, 30*(1), 1–19.

Aziz, A. A., Sheikh, S. N. S., Yusof, K. M., Udin, A., & Yatim, J. M. (2012). Developing a structural model of assessing students' knowledge-attitudes towards sustainability. *Procedia-Social and Behavioral Sciences, 56*(1), 513–522.

Bakar, A. Y. A. (2017). Developing gifted and talented education program: The Malaysian experience. *Creative Education, 8*(1), 1–11.

Barth, M., & Rieckmann, M. (2012). Academic staff development as a catalyst for curriculum change towards education for sustainable development: An output perspective. *Journal of Cleaner Production, 26*(1), 28–36.

Bell, B. (1998). Teachers' development in science education. In B. J. Fraser & K. G. Tobin (Eds.), *International handbook of science education* (pp. 681–694). Dordrecht: Kluwer Academic Publishers.

Biedenweg, K., Monroe, M. C., & Oxarart, A. (2013). The importance of teaching ethics of sustainability. *International Journal of Sustainability in Higher Education, 14*(1), 6–14.

Blum, C., Bunke, D., Hungsberg, M., Roelofs, E., Joas, A., Joas, R., et al. (2017). The concept of sustainable chemistry: Key drivers for the transition towards sustainable development. *Sustainable Chemistry and Pharmacy, 5*(1), 94–104.

Borg, C., Gericke, N., Hoglund, H. O., & Bergman, E. (2012). The barriers encountered by teachers implementing education for sustainable development: Discipline bound differences and teaching traditions. *Research in Science and Technological Education, 30*(2), 185–207.

Bradley, J. D. (2005). Chemistry education for development. *Chemical Education International, 6*(1), 1–6.

Burmeister, M., & Eilks, I. (2012). An example of learning about plastics and their evaluation as a contribution to education for sustainable development in secondary school chemistry teaching. *Chemistry Education Research and Practice, 13*(2), 93–102.

Burmeister, M., & Eilks, I. (2013). An understanding of sustainability and education for sustainable development among German student teachers and trainee teachers of chemistry. *Science Education International, 24*(2), 167–194.

Burmeister, M., Rauch, F., & Eilks, I. (2012). Education for sustainable development (ESD) and chemistry education. *Chemistry Education Research and Practice, 13*(2), 59–68.

Cebrian, G., Grace, M., & Humphris, D. (2015). Academic staff engagement in education for sustainable development. *Journal of Cleaner Production, 106*(1), 79–86.

Chakraborty, A., Singh, M. P., & Roy, M. (2016). Barriers in restructuring university curriculum for a sustainable future. *Annual Research Journal of SCMS, Pune, 4*(1), 67–79.

Chen, C. J., Gregoire, M., Arendt, S. W., & Shelley, M. C. (2011). College and university dining services administrators' intention to adopt sustainable practices: Results from US institutions. *International Journal of Sustainability in Higher Education, 12*(2), 145–162.

Choy, E. A., Basan, N. A., Hussin, R., Zei, L. H., Wahab, A. A., Mustapa, F. L., et al. (2017). Sustainable campus initiative: The level of knowledge, awareness and sustainable practice among staff and students at the National University of Malaysia. *Journal of Global Business and Social Entrepreneurship, 3*(6), 65–73.

Cotton, D., Bailey, I., Warren, M., & Bissell, S. (2009). Revolutions and second best solutions: Education for sustainable development in higher education. *Studies in Higher Education, 34*(7), 719–733.

Crespo, B., Alvarez, C. M., Arce, M. E., Cuevas, M., & Miguez, J. L. (2017). The sustainable development goals: An experience on higher education. *Sustainability, 9*(1353), 1–15.

Darwish, M. M., Agnello, M. F., & Burgess, R. (June 2010). Incorporating sustainable development and environmental ethics into construction engineering education. In *Eighth LACCEI Latin American and Caribbean Conference for Engineering and Technology* (pp. 1-4). Arequipa, Peru.

Davey, I. (2012). Roles of awareness and intention in determining levels of environmentally positive action: A review of studies. *Journal of Administration and Governance, 7*(1), 23–42.

Disterheft, A., Caeiro, S. S. F. D. S., Ramos, M. R., & Azeiteiro, U. M. D. M. (2012). Environmental Management Systems (EMS) implementation processes and practices in European higher education institutions-topdown versus participatory approaches. *Journal of Cleaner Production, 31*(1), 80–90.

Dwyer, W. O., Leeming, F. C., Cobern, M. K., Porter, B. E., & Jackson, J. M. (1993). Critical review of behavioral interventions to preserve the environment: Research since 1980. *Environment and Behavior, 25*(5), 275–321.

Economic Planning Unit. (2015). *The Eleventh Malaysia Plan 2016–2020*. Kuala Lumpur: Percetakan Nasional Malaysia Berhad (PNMB).

Egarievwe, S. U. (2015). Vertical education enhancement: A model for enhancing STEM education and research. *Procedia Social and Behavioral Sciences, 177*(1), 336–344.

Eissen, M. (2012). Sustainable production of chemicals-an educational perspective. *Chemistry Education Research and Practice, 13*(2), 103–111.

Emanuel, R., & Adams, J. (2011). College students' perceptions of campus sustainability. *International Journal of Sustainability in Higher Education, 12*(1), 79–92.

Fabbrizzi, S., Maggino, F., Marinelli, N., Menghini, S., & Ricci, C. (2016). Sustainability and well-being: The perception of younger generations and their expectations. *Agriculture and Agricultural Science Procedia, 8*(1), 592–601.

Fah, L. Y., & Sirisena, A. (2014). Relationships between the knowledge, attitudes, and behaviour dimensions of environmental literacy: A structural equation modeling approach using smartpls. *Journal for Educational Thinkers, 5*(1), 119–144.

Figueiro, P. S., & Raufflet, E. (2015). Sustainability in higher education: A systematic review with focus on management education. *Journal of Cleaner Production, 106*(1), 22–33.

Filho, W. L. (2000). Dealing with misconceptions on the concept of sustainability. *International Journal of Sustainability in Higher Education, 1*(1), 9–19.

Flamm, B. J. (2006). *Environmental knowledge, environmental attitudes, and vehicle ownership and use* (Ph.D. thesis). Graduate Division, University of California, Berkeley.

Foo, K. Y. (2013). A vision on the role of environmental higher education contributing to the sustainable development in Malaysia. *Journal of Cleaner Production, 61*(1), 6–12.

Gomani, M. S. (2010). A case study on initiatives in the current use of integrating education for sustainable development in TVET in Malawi. In R. Dubois, K. Balgobin, M. S. Gomani, J. K. Kelemba, G. S. Konayuma, M. L. Phiri, & J. W. Simiyu (Eds.), *Integrating sustainable development in technical, vocational education, and training* (pp. 55–67). Bonn: UNESCO-UNEVOC International Centre for Technical and Vocational Education and Training.

Gusti, A. (2016). The relationship of knowledge, attitudes, and behavioral intentions of sustainable waste management on primary school students in city of Padang, Indonesia. *International Journal of Applied Environmental Sciences, 11*(5), 1323–1332.

Haan, G. D. (2006). The BLK'21' programme in Germany: A 'Gestaltungskompetenz'-based model for education for sustainable development. *Environmental Education Journal, 12*(1), 19–32.

Haan, G. D., Bormann, I., & Leicht, A. (2010). The midway point of the UN decade of education for sustainable development: Current research and practice in ESD. *International Review of Education, 56* (2–3), 199–206.

Hargreaves, A. (1994). *Changing teachers, changing times: Teachers' work and culture in the postmodern age*. Toronto: OISE Press.

Hassan, A., Noordin, T. A., & Sulaiman, S. (2010). The status on the level of environmental awareness in the concept of sustainable development amongst secondary school students. *Procedia Social and Behavioral Sciences, 2*(2), 1276–1280.

Hasan, M. N. H. (1998). Indicators of sustainable development: The malaysian perspective. In *Proceedings of Regional Dialogue on Geo-Indicators for Sustainable Development, Institute for Environment and Development (LESTARI)*, Bangi, Malaysia, (pp. 1–16).

Hayles, C. S., & Holdsworth, S. E. (2005). Constructing stimulus: Teaching sustainability to engender change. In *Fabricating Sustainability: 39th Annual Conference of the Architectural Science Association*.

Heaton, A., Hodgson, S., Overton, T., & Powell, R. (2006). The challenge to develop CFC (chlorofluorocarbon) replacements: A problem-based learning case study in green chemistry. *Chemistry Education Research and Practice, 7*(4), 280–287.

Hegarty, K. (2008). Shaping the self to sustain the other: Mapping impacts of academic identity in education for sustainability. *Environmental Education Research, 14*(6), 681–692.

Hines, J. M., Hungerford, H. R., & Tomera, A. N. (1987). Analysis and synthesis of research on responsible environmental behavior: A meta-analysis. *The Journal of Environmental Education, 18*(2), 1–8.

Hoang, T. T. P., & Kato, T. (2016). Measuring the effect of environmental education for sustainable development at elementary schools: A case study in Da Nang city, Vietnam. *Sustainable Environment Research, 26*(6), 274–286.

Hofman, M. (2015). What is an education for sustainable development supposed to achieve-a question of what, how and why. *Journal of Education for Sustainable Development, 9*(2), 213–228.

Hofstein, A., Eilks, I., & Bybee, R. (2011). Societal issues and their importance for contemporary science education: A pedagogical justification and the state of the art in Israel, Germany, and the USA. *International Journal of Science and Mathematics Education, 9*(6), 1459–1483.

Holbrook, J., & Rannikmae, M. (2009). The meaning of scientific literacy. *International Journal of Environmental and Science Education, 4*(3), 275–288.

Idros, S. N. S. (2006). Exploring environmental behaviours, attitudes and knowledge among university students: Positioning the concept of sustainable development within Malaysian education. *Journal of Science and Mathematics Education, 29*(1), 79–97.

Isa, N. K. M. (2016). Sustainable campus and academic staffs' awareness and behaviour in Malaysia's institutions of higher learning: a case study of UPSI. *Geografia-Malaysian Journal of Society and Space, 12*(6), 89–99.

Iyer, L. S. (2018). Knowledge, attitude and behaviour (KAB) of student community towards electronic waste: A case study. *Indian Journal of Science and Technology, 11*(10), 1–10.

Johannesson, I. A., Norodahl, K., Oskarsdottir, G., Palsdottir, A., & Petursdottir, B. (2011). Curriculum analysis and education for sustainable development in Iceland. *Environmental Education Research, 17*(3), 375–391.

Jorge, M. L., Madueno, J. H., Cejas, M. Y. C., & Pena, F. J. A. (2015). An approach to the implementation of sustainability practices in Spanish universities. *Journal of Cleaner Production, 106*(1), 34–44.

Juntunen, M., & Aksela, M. (2014). Education for sustainable development in chemistry: Challenges, possibilities and pedagogical models in Finland and elsewhere. *Chemistry Education Research and Practice, 15*(4), 488–500.

Kagawa, F. (2007). Dissonance in students' perceptions of sustainable development and sustainability: Implications for curriculum change. *International Journal of Sustainability in Higher Education, 8*(3), 317–338.

Kanapahty, S., Lee, K. E., Mokhtar, M., Zakaria, S. Z. S., Sivapalan, S., & Zahidi, A. M. 2018. The integration of sustainable development concept in Chemistry curriculum: A conceptual framework for the case of Pusat PERMAT Apintar$^{TM}$ Negara. In *7th World Engineering Education Forum, (WEEF 2017)* (pp. 303–308).

Kanapahty, S., Lee, K. E., Sivapalan, S., Mokhtar, M., Zakaria, S. Z. S., & Zahidi, A. M. (2019a). Sustainable development concept in the chemistry curriculum: An exploration of foundation students' perspective. *International Journal of Sustainability in Higher Education, 20*(1), 2–22.

Kanapahty, S., Lee, K. E., Mokhtar, M., Zakaria, S. Z. S., Sivapalan, S., & Zahidi, A. M. (2019b). Sustainable chemistry teaching at the pre-university level: Barriers and opportunities for university educators. *International Journal of Sustainability in Higher Education, 20*(4), 784–802.

Karabell, Z. (1998). *What's college for? The struggle to define American higher education*. Scranton: Harper Collins Publishers.

Karpudewan, M., Hj Ismail, Z., & Mohamed, N. (2009). The integration of green chemistry experiments with sustainable development concepts in pre-service teachers' curriculum: Experiences from Malaysia. *International Journal of Sustainability in Higher Education, 10*(2), 118–135.

Karpudewan, M., Hj Ismail, Z., & Mohamed, N. (2011a). Greening a chemistry teaching methods course at the School of Educational Studies, Universiti Sains Malaysia. *Journal of Education for Sustainable Development, 5*(2), 197–214.

Karpudewan, M., Ismail, Z., & Roth, W. M. (2012). Ensuring sustainability of tomorrow through green chemistry integrated with sustainable development concepts (SDCs). *Chemistry Education Research and Practice, 13*(2), 120–127.

Karpudewan, M., Ismail, Z. H., & Mohamed, N. (2011b). Green chemistry: Educating prospective science teachers in education for sustainable development at School of Educational Studies, Universiti Sains Malaysia. *Journal of Social Sciences, 7*(1), 42–50.

Khataybeh, A. M., Subbarini, M., & Shurman, S. (2010). Education for sustainable development, an international perspective. *Procedia Social and Behavioral Sciences, 5*(1), 599–603.

Kollmuss, A., & Agyeman, J. (2002). Mind the gap: Why do people act environmentally and what are the barriers to pro-environmental behavior? *Environmental Education Research, 8*(3), 239–260.

Labodova, A., Lapcik, V., Kodymova, J., Turjak, J., & Pivko, M. (2014). Sustainability teaching at VSB: Technical university of Ostrava. *Journal of Cleaner Production, 62*(1), 128–133.

Landorf, H., Doscher, S., & Rocco, T. (2008). Education for sustainable human development: Towards a definition. *Theory and Research in Education, 6*(2), 221–236.

Laor, P., Suma, Y., Keawdounglek, V., Hongtong, A., Apidechkul, T., & Pasukphun, N. (2018). Knowledge, attitude and practice of municipal solid waste management among highland residents in Northern Thailand. *Journal of Health Research, 32*(2), 123–131.

Laroche, M., Bergeron, J., & Barbaro-Forleo, G. (2001). Targeting consumers who are willing to pay more for environmentally friendly products. *Journal of Consumer Marketing, 18*(6), 503–520.

Launiala, A. (2009). How much can a KAP survey tell us about people's knowledge, attitudes and practices? Some observations from medical anthropology research on malaria in pregnancy in Malawi. *Anthropology Matters, 11*(1), 1–13.

Le, X. Q., Le, H. H., Vu, D. C., Nguyen, H. N., Nguyen, T. T. A., & Vu, T. H. N. (2015). Integrated Science, Technology, Engineering and Mathematics (STEM) education through active experience of designing technical toys in Vietnamese schools. *British Journal of Education, Society & Behavioural Science, 11*(2), 1–12.

Lenglet, F., Fadeeva, Z., & Mochizuki, Y. (2010). ESD promises and challenges: Increasing its relevance. *Global Environment Research, 14*(2), 93–100.

Lozano, R. (2006). Incorporation and institutionalization of SD into universities: Breaking through barriers to change. *Journal of Cleaner Production, 14*(9), 787–796.

Mahat, H., Ahmad, S., Ngah, M. S. Y. C., & Ali, N. (2014). Sustainable development education—The awareness relationship between students and parents. *Malaysian Journal of Society and Space, 10*(5), 71–84.

Mahat, H., & Idrus, S. (2016). Education for sustainable development in Malaysia: A study of teacher and student awareness. *Malaysian Journal of Society and Space, 12*(6), 77–88.

Mahat, H., Ngah, M. S. Y. C., & Idrus, S. (2013a). Sustainable development education awareness among student through sustainable school programs. In *The 4th Geography and Environment Conference* (Vol. 1, no. 2, pp. 44–58).

Mahat, H., Ngah, M. S. Y. C., & Idrus, S. (2013b). A study on the importance of teacher knowledge in implementing sustainable school programs in Malaysia. *Journal of Social Science and Humanity, 5*(2), 75–92.

Mahmud, M. H. B., & Siarap, K. B. H. (2013). H1N1 prevention campaign: A study of knowledge, attitudes and practices of the residents in northeast Penang. *Journal of Communication, 29*(1), 127–140.

Mandler, D., Mamlok-Naaman, R., Blonder, R., Yayon, M., & Hofstein, A. (2012). High school chemistry teaching through environmentally oriented curricula. *Chemistry Education Research and Practice, 13*(2), 80–92.

Manitoba Education and Training. (2000). *Education for a sustainable future: A resource for curriculum developers, teachers, and administrators*. Manitoba: Minister of Education and Training.

Marginson, S., Tytler, R., Freeman, B., & Roberts, K. (2013). STEM: Country comparisons: International comparisons of science, technology, engineering and mathematics (STEM) education. Final report.

Marks, R., & Eilks, I. (2009). Promoting scientific literacy using a sociocritical and problem oriented approach to chemistry teaching: Concept, examples, and experiences. *International Journal of Environmental and Science Education, 4*(3), 231–245.

Matsuura, K. (2004). Why education and public awareness are indispensable for a sustainable future. In *Proceedings of High-Level International Conference on Education for Sustainable Development* (pp. 27–31).

McDonald, C. (2006). Moving forward on educating for sustainability in Manitoba. *Journal of Cleaner Production, 14*(1), 10–16.

McKeown, R., & Hopkins, C. (2007). Moving beyond the EE and ESD disciplinary debate in formal education. *Journal of Education for Sustainable Development, 1*(1), 17–26.

McKeown, R., Hopkins, C. A., Rizi, R., & Chrystalbridge, M. (2002). *Education for sustainable development toolkit*. University of Tennessee, Knoxville: Energy, Environment and Resources Center.

Meerah, T. S. M., Halim, L., & Nadeson, T. (2010). Environmental citizenship: What level of knowledge, attitude, skill and participation the students own? *Procedia-Social and Behavioral Sciences, 2* (2), 5715–5719.

Mehlmann, M., McLaren, N., & Pometun, O. (2010). Learning to live sustainably. *Global Environmental Research, 15*(1), 177–186.

Millar, R. (2008). Taking scientific literacy seriously as a curriculum aim. In *Asia-Pacific forum on science learning and teaching* (Vol. 9, no. 2, pp. 1–8). The Education University of Hong Kong: Department of Science and Environmental Studies.

Minghat, A. D., & Yasin, R. M. (2010). A sustainable framework for technical and vocational education in Malaysia. *Procedia Social and Behavioral Sciences, 9*(1), 1233–1237.

Ministry of Education Malaysia. (2013). *Malaysia Education Blueprint 2013–2025 (Preschool to Post-Secondary Education).* Putrajaya: Ministry of Education Malaysia.

Mintz, K., Talesnick, M., Amadei, B., & Tal, T. (2013). Integrating sustainable development into a service learning engineering course. *Journal of Professional Issues in Engineering Education and Practice, 140*(1), 1–11.

Moore, J. (2005). Barriers and pathways to creating sustainability education programs: Policy, rhetoric and reality. *Environmental Education Research, 11*(5), 537–555.

Ndinechi, M.C., & Okafor, K. C. (2016). STEM education: A tool for sustainable national capacity building in a digital economy. In *1st International Conference of Federal University of Technology*

*Owerri—The Centre for Continuing Education (FUTO-CCE)*, May 16-19, 2016, (pp. 1–10) Nigeria.

Ngah, K., Mustaffa, J., Zakaria, Z., Noordin, N., & Sawal, M. Z. H. M. (2011). Formulation of Agenda 21 process indicators for Malaysia. *Journal of Management and Sustainability, 1*(1), 82–89.

Obama, B. (2009). Educate to innovate press conference, viewed 15 December 2016. http://www.whitehouse.gov/issues/education/educate-innovate.

Osborne, J., & Dillon, J. (2008). *Science education in Europe: Critical reflections*. London: The Nuffield Foundation.

Papadimitriou, V. (2004). Prospective primary teachers' understanding of climate change, greenhouse effect, and ozone layer depletion. *Journal of Science Education and Technology, 13*(2), 299–307.

Paulus, S. C. (1996). Exploring a pluralist understanding of learning for sustainability and its implications for outdoor education practice. *Journal of Adventure Education and Outdoor Learning, 16*(2), 117–130.

Peter, C. J., Libunao, W. H., & Latif, A. A. (2016). Extent of education for sustainable development (ESD) integration in Malaysian community colleges. *Journal of Technical Education and Training, 8*(1), 1–13.

Pigozzi, M. J. (2010). Implementing the UN decade of education for sustainable development (DESD): Achievements, open questions and strategies for the way forward. *International Review of Education, 56*(2–3), 255–269.

Pitt, J. (2009). Blurring the boundaries–STEM education and education for sustainable development. *Design and Technology Education: An International Journal, 14*(1), 1360–1431.

Pudin, S., Tagi, K., & Periasamy, A. (2004). Environmental education in Malaysia and Japan: A comparative assessment, viewed 24 April 2018. http://www.ceeindia.org/esf/download/paper20.pdf.

Pusat PERMATApintarTM Negara. (2018). ASASIpintar (Pra-Univ), viewed 1st Mac 2018. http://www.ukm.my/permatapintar/asasipintar/.

Ramsey, C. E., & Rickson, R. E. (1976). Environmental knowledge and attitudes. *The Journal of Environmental Education, 8*(1), 10–18.

Reza, M. I. H. (2016). Sustainability in higher education: Perspectives of Malaysian higher education system. *Sage Open, 6*(3), 1–9.

Rocard, M., Csermely, P., Jorde, D., Lenzen, D., Henriksson, H. W., & Hemmo, V. (2007). *Science education now: A renewed pedagogy for the future of Europe*. Belgium: European Commission.

Ryan, A., & Cotton, D. (2013). Times of change: Shifting pedagogy and curricula for future sustainability. In S. Sterling, L. Maxey, & H. Luna (Eds.), *The sustainable university: Progress and prospects* (pp. 15–167). Oxon: Routledge.

Sadler, T., & Dawson, V. (2012). Socio-scientific issues in science education: Contexts for the promotion of key learning outcomes. In B. Fraser, K. Tobin, & C. McRobbie (Eds.), *Second international handbook of science education* (pp. 799–809). Dordrecht: Springer.

Sahin, E., Ertepinar, H., & Teksoz, G. (2012). University students' behaviors pertaining to sustainability: A structural equation model with sustainability related attributes. *International Journal of Environmental and Science Education, 7*(3), 459–478.

Said, A. M., Yahaya, N., & Ahmadun, F. I. R. (2007). Environmental comprehension and participation of Malaysian secondary school students. *Environmental Education Research, 13*(1), 17–31.

Shari, Z., & Jaafar, M. F. Z. (2006). Towards a holistic sustainable architectural education in Malaysia. *International Journal on Sustainable Tropical Design Research and Practice, 1*(1), 57–65.

Sheikh, S. N. S., Aziz, A. A., & Yusof, K. M. (2012). Perception on sustainable development among new first year engineering undergraduates. *Procedia-Social and Behavioral Sciences, 56*(1), 530–536.

Sipos, Y., Battisti, B., & Grimm, K. (2008). Achieving transformative sustainability learning: Engaging head, hands and heart. *International Journal of Sustainability in Higher Education, 9*(1), 68–86.

Sivapalan, S. (2016). Engineering education for sustainable development in Malaysia: Student stakeholder's perspectives on the integration of holistic sustainability competences within undergraduate engineering programmes. In W. L. Filho & L. Brandli (Eds.), *Engaging stakeholders in education for sustainable development at university level* (pp. 263–285). Switzerland: Springer International Publishing.

Sjostrom, J., & Stenborg, E. (2014). Teaching and learning for critical scientific literacy: Communicating knowledge uncertainties, actors' interplay and various discourses about chemicals. In I. Eilks, S. Markic, & B. Ralle (Eds.), *Science education research and education for sustainable development* (pp. 37–47). Germany: Shaker Verlag.

Sterling, S. (1996). Education in change. In J. Huckle, R. Stephen, Sterling, & S. Sterling (Eds.), *Education for sustainability* (pp. 18–39). London: Earthscan Publication.

Sterling, S. (2001). *Sustainable education: Re-Visioning learning and change. Schumacher Briefings*. Seaton Road, Bristol, (BS1 6XN), England: Schumacher UK CREATE Environment Centre, (6 pounds).

Sterling, S. (2004). Higher education, sustainability and the role of systematic learning. In P. B. Corcoran & A. E. J. Wals (Eds.), *Higher education and the challenge of sustainability: Problematics, promise and practice* (pp. 49–70). Dordrecht: Kluwer Academic Publisher.

Strauss, B. H. (1996). *The class of 2000 report: Environmental education, practices, and activism on campus*. New York: Environment Program, Nathan Cummings Foundation.

Suryawanshi, K., & Narkhede, S. (2015). Green ICT for sustainable development: A higher education perspective. *Procedia Computer Science, 70*(1), 701–707.

Tal, T., Kali, Y., Magid, S., & Madhok, J. J. (2011). Enhancing the authenticity of a web-based module for teaching simple inheritance. In T. D. Sadler (Ed.), *Socio scientific issues in the classroom* (pp. 11–38). Dordrecht: Springer.

Tal, T., & Kedmi, Y. (2006). Teaching socio scientific issues: Classroom culture and students' performances. *Cultural Studies of Science Education, 1*(4), 615–644.

Taylor, N., Nathan, S., & Coll, R. K. (2003). Education for sustainability in regional New South Wales, Australia: An exploratory study of some teachers' perceptions. *International Research in Geographical and Environmental Education, 12*(4), 291–311.

Terenzini, P. T., Cabrera, A. F., Colbeck, C. L., Parente, J. M., & Bjorklund, S. A. (2001). Collaborative learning vs. lecture/discussion: Students' reported learning gains. *Journal of Engineering Education, 90*(1), 123–130.

The Cloud Institute for Sustainability Education. (2019). viewed 1 Mac 2019. https://cloudinstitute.org/.

Thomas, I. (2004). Sustainability in tertiary curricula: What is stopping it happening. *International Journal of Sustainability in Higher Education, 5*(1), 33–47.

UNESCO. (2005). United Nations decade of education for sustainable development (2005–2014). In *International Implementation Scheme*. Paris: UNESCO.

UNESCO. (2011). Country reports on education for sustainable development. In *Centred on the Five Cluster Countries of UNESCO Office, Jakarta*. Jakarta: UNESCO.

UNESCO. (2012). Shaping the education of tomorrow. In *2012 Report on the United Nation Decade of Education for Sustainable Development, Abridged*. Paris: UNESCO.

UNESCO. (2013). *Proposal for a global action programme on education for sustainable development as follow-up to the united nations decade of education for sustainable development after 2014*. France: UNESCO.

UNESCO. (2015). *Education for sustainable development policy*. Guyana: UNESCO.

United Nations. (2015). Transforming our world: The 2030 agenda for sustainable development, viewed 30 November 2016. https://sustainabledevelopment.un.org/content/documents/2125030%20Agenda%20for%20Sustainable%20Development%20web.pdf.

Vandamme, E. (2009). *Concepts and challenges in the use of knowledge attitude practice surveys: Literature review*. Belgium: Institute of Tropical Medicine.

Vasiliki, L. V., & Maria, L. (2015). The three pillars of sustainability. In G. Goniadis & M. Lampridi (Eds.), *Introduction to sustainable development* (pp. 25–27). Greece: International Hellenic University.

Velazquez, L., Munguia, N., & Sanchez, M. (2005). Deterring sustainability in higher education institutions: An appraisal of the factors which influence sustainability in higher education institutions. *International Journal of Sustainability in Higher Education, 6* (4), 383–391.

Venkataraman, B. (2009). Education for sustainable development. *Environment: Science and Policy for Sustainable Development, 51* (2), 8–10.

Veugelers, W. (2000). Different ways of teaching values. *Educational Review, 52*(1), 37–46.

Viebahn, P. (2002). An environmental management model for universities: From environmental guidelines to staff involvement. *Journal of Cleaner Production, 10*(1), 3–12.

Wals, A. E. (2007). *Social learning towards a sustainable World: Principles, perspectives, and praxis*. Netherlands: Wageningen Academic Publishers.

Wals, A. E. (2010). Mirroring, gestaltswitching and transformative social learning: Stepping stones for developing sustainability competence. *International Journal of Sustainability in Higher Education, 11*(4), 380–390.

Wan Nur'ashiqin, W. M., Er, A. C., Noraziah, A., Novel, L. H., Saadiah, H. S., & Buang, A. (2011). Diagnosing knowledge, attitudes and practices for a sustainable campus. *World Applied Sciences Journal, 13*(13), 93-98.

Ware, S. A. (2001). Teaching chemistry from a societal perspective. *Pure and Applied Chemistry, 73*(7), 1209–1214.

Wheeler, K. (2000). Sustainability from five perspectives. In K. A. Wheeler & A. P. Bijur (Eds.), *Education for a sustainable future* (pp. 2–6). New York: Springer.

Williams, J. (2011). STEM education: Proceed with caution. *Design and Technology Education: An International Journal, 16*(1), 26–35.

Wilson, S. (2012). Drivers and blockers: Embedding education for sustainability (EfS) in primary teacher education. *Australian Journal of Environmental Education, 28*(1), 42–56.

Wright, T. S. A. (2006). Giving "teeth" to a university sustainability policy: A Delphi study at Dalhousie University. *Journal of Cleaner Production, 14*(1), 761–768.

Yassin, S. F. M., Ishak, N. M., Yunus, M. M., & Majid, R. A. (2012). The identification of gifted and talented students. *Procedia Social and Behavioral Sciences, 55*(1), 585–593.

Yencken, D., Fien, J., & Sykes, H. (2000). *Environment, education and society in the Asia-Pacific: Local traditions and global discourses*. London: Routledge.

Yin, T. C., Huang, C. C., & Kawata, C. (2002). The effects of different environmental education programs on the environmental behavior of seventh grade students and related factors. *Journal of Environmental Health, 64*(7), 24–29.

Yuan, X., & Zuo, J. (2013). A critical assessment of the higher education for sustainable development from students' perspectives–a Chinese study. *Journal of Cleaner Production, 48*(1), 108–115.

Zamhari, S. K., & Perumal, C. (2016). Challenges and strategies towards a sustainable community. *Geografia-Malaysia Journal of Society and Space, 12*(12), 10–24.

# Mainstreaming, Institutionalizing and Translating Sustainable Development Goals into Non-governmental Organization's Programs

Mohamad Muhyiddin Hassan, Khai Ern Lee, and Mazlin Mokhtar

**Abstract**

After the introduction of the Sustainable Development Goals (SDGs) in 2015, non-governmental organizations (NGOs) become a prime mover of SDGs in representing the diverse range of organizational interests and broadening the social aspects of civil society beyond the other sectors. The existence of NGOs not only represents the voice of the civil society but also fills the gaps when the constituencies of the government and industry sectors are limited. However, the capacities of NGOs are influenced by their uncertainties, especially in the transition from single-sector approach to cross-sector approach to increase trade-offs between the SDGs. Thus, mainstreaming and institutionalizing SDGs are very important in translating SDGs into NGOs' program implementation. In this chapter, an international NGO based in Malaysia, the Global Environment Centre (GEC), is taken as a case study whereby a framework has been proposed, consisting of three strategies to streamline NGOs' programs toward achieving the SDGs. The framework is instrumental to guide NGOs to implement SDGs through bottom-up approach by translating every SDGs into action-oriented programs, forging hybrid governance for cooperation among NGOs' partner institutions and making social value the essence of fostering environmental citizenship. As NGOs have unmeasurable capacities, this strategy can help NGOs in mainstreaming, institutionalizing and translating SDGs into their projects as a measurement of project performance and can be standardized despite diverse project scope and eventually help to achieve the SDGs at large.

**Keywords**

Non-governmental organization • Sustainable Development Goals • Organization • Institution • Governance

## 1 Sustainable Development

The term sustainable development was first expressed by the International Union for Conservation of Nature (IUCN) in its reporting, "World Conservation Strategy" in 1980 (Hopwood et al. 2005). The concept of sustainable development became more prominent as a result of the 1987 Brundtland Report or "Our Common Future" which was classified as a classic definition whereby a development that meets the needs of the present generation without compromising the ability of future generations to meet their own needs (Brundtland 1987; Giddings et al. 2002). However, the concept of sustainable development was initially challenged by its less robust and vague theory. Although the concept of sustainable development emphasizes environmental protection, deep ecologists reject the definition because it is largely viewed from the standpoint of human rather than the environmental interest (Giddings et al. 2002). The definition of Brundtland's need to "meet the needs of humanity both present and future" has left a strong impression that sustainable development is the embodiment of the development of every human desire, meaning that such seemingly simple development is inherently mistaken as the hidden concept behind the concept of sustainable development (Redclift 2005). Even the ambiguity of the concept of sustainable development is increasingly complicated when the term used is a form of syllogism to understand on human needs. This refers to the verse "needs of the present and future" that is actually trying to describe the different form of human needs which is beyond control or change from time to time and at different places of the world (Redclift 2005). The changing nature of human needs as a general statement refers to the

M. M. Hassan · K. E. Lee (✉) · M. Mokhtar
Institute for Environment and Development (LESTARI), Universiti Kebangsaan Malaysia, Bangi, 43600, Selangor, Malaysia
e-mail: khaiernlee@ukm.edu.my


K. E. Lee (ed.), *Concepts and Approaches for Sustainability Management*, Advances in Science, Technology & Innovation, https://doi.org/10.1007/978-3-030-34568-6_6

evolving human needs of the ages, while the changing nature of human needs as a specific expression refers to different human needs according to local cultural patterns.

Daly (1990) described the term sustainable development as an oxymoron because sustainable development is a combination of phrases that contradict each other and produce rhetorical effects. Ambiguity of sustainable development can have the effect of forming a political rhetoric and may even result in demagogy. Political rhetoric gives the freedom of using the of term sustainable development as catchphrase that is seen as more fashionable and up to date (*de rigueur*) by politicians and businessmen, whereas demagogy is a political issue that can ignite the irrationality of the people in examining the fundamental principles and practices of a development system because they do not have a clear understanding of a developmental need (Hopwood et al. 2005). Although Gro Harlem Brundtland has embraced the concept of sustainable development in her political rhetoric, she also provided the political platform for sustainable development to grow as an early stage in developing sustainable development concept on a regular basis (Daly 1990). As a result, the concept of sustainable development has become a global agenda for two reasons, namely evidence of global concern for environmental destruction and the worst record of post-World War II development (Kemp et al. 2005). Although widespread, the absence of a clear theoretical and analytical framework of sustainable development concept makes it difficult to determine new policies in development that are supposed to foster the love for environmental and give meaning to society (Lele 1991). The absence of semantic description and concept clearly prevents some debates from producing the results of what constitutes sustainable development (Lele 1991). The ambiguity and misunderstanding of semantic description and the endless generation of concepts have caused the concept of sustainable development to have no fixed meaning until it has been left a paradox (Tarlock 2001).

Sustainable development has continued to move forward with the adoption of a holistic approach of a combination of socioeconomic environmental dimensions since the Earth Summit in Rio De Janeiro in 1992 (Grybaitė and Tvaronavičienė 2008). The approach of sustainable development dimensions is a reflection of a number of research approaches and descriptions of human life and the world around them that are dominated by different disciplines of knowledge (Giddings et al. 2002). The dominance of the diversity of knowledge disciplines in adopting sustainable development dimensions' approach requires one governance to elaborate the concept of sustainable development in order to be practical. The complexity of sustainable development can be classified into five areas, namely differentness, trade-offs and uncertainty, ambiguity and diversity, interconnections and integration and normative principle (Hezri 2016; Kemp et al. 2005).

According to Kemp et al. (2005), four key elements of governance need to be integrated into the adaptive change of the social institutional process toward the complexity of sustainable development in order to transform the concept of sustainable development into a practical or action-oriented one. First, policy integration refers to the coordination of specialized jurisdictions to be more flexible as separate legal practices only help to resolve certain issues but do not address issues across sector boundaries. Second, policy instruments (objectives, criteria, alternatives and common indicators) refer to structured methods of planning and implementation that have action and progress toward sustainability. Third, information and incentives for practical implementation refer to the foundations that can guide the sustainability-based decision-making process. Fourth, the program for system innovation refers to a technical component that emphasizes the fundamental changes in the system of provision of goods in order to utilize different resources, knowledge and practices on a sustainable basis. These four key elements of governance not only provide a platform for adopting sustainable development dimension approach as the goal of transferring sustainable development concept toward practical development, but also provide a platform for translating sustainable development concept into goal-setting features.

## 2 Sustainable Development Goals

The Millennium Development Goals (MDGs) are a set of goals set during the Millennium Conference in 2000 through the UN 55/2 resolution or the UN Millennium Declaration (Hulme 2009). MDGs make history in creating effective ways for the global transition to promote global awareness, political accountability, better metrics, social feedback and public pressure to achieve social priorities worldwide by focusing public attention on its eight goals (Sachs 2012). Today, in the era of post-Rio + 20, UN member states are implementing the latest UN Goals, Sustainable Development Goals (SDGs), to replace MDGs that expired in 2015 after fifteen years of global change (Sachs 2012).

SDGs are the second global policy instruments of sustainable development after the end of the implementation of MDGs. Chronologically, the concept of sustainable development began with the movement of environmentalism which was then further integrated with another two pillars, namely social and economic issues that form an effort in the form of goal setting (Tarlock 2001). The enrichment of policy instruments (one of the key elements of governance), such as the practical use of sustainability indicators (Hezri and Dovers 2006) in sustainable development, led to the emergence of the idea of goal setting.

From an idea without institutions (Tarlock 2001), institutions are increasingly expanding by introducing several indicators of sustainable development as one of the key policy instruments (Grybaite and Tvaronavičiene 2008). Six recognized international institutions have sustainability indicators, such as the Directorate-General of the European Commission (Eurostat), the UN, the European Environment Agency, the Organisation for Economic Co-operation and Development (OECD), the Statistical Indicators Benchmarking the Information Society (SIBIS) and the Directorate-General for Enterprise and Industry (DG ENTR). However, the system of indicators developed from each of these institutions is different, and they do not pay attention to the overlap and interdependence of the indicators, as such sustainable development is integrated, comprehensive and inclusive. Such weaknesses can be seen in the implementation of MDGs with no more inclusive indicators as several dimensions need to be addressed that are not involved, such as human rights and economic growth, while the complexity of dimensions is not included, such as gender equality and quality of education (Vandemoortele 2011). In addition, the implementation of SDGs requires a strong characterization that enhances not only the usefulness of its indicators, but also the need for SDGs indicators themselves to be relevant in enhancing the improvement of every aspect of sustainable development.

## 2.1 SDGs Implementation

The Sustainable Development Goals (SDGs) are the latest United Nations (UN) global initiative known as the "Agenda 2030: Transformation of Our World" and themed "Leave No One Behind" (Klasen and Fleurbaey 2018). The SDGs were launched during the UN General Assembly held on September 25, 2015, and 193 UN General Assembly members approved UN resolution 70/1 to set the implementation of the SDGs in place of the Millennium Development Goals (MDGs) that expired in 2015, after fifteen years of improving the economy of developing countries through its eight goals (Sachs 2012; Griggs et al. 2013). The implementation of the SDGs was more challenging following the end of the MDGs and left the remaining global issues to be eradicated (United Nations 2015):

(i) The continuing existence of gender inequality;
(ii) Large gaps between rich and poor households between urban and rural areas;
(iii) Climate change and environmental degradation undermine progress achieved and the poor most affected;
(iv) Humanitarian conflict is the greatest threat to development; and
(v) Millions of the poor live under extreme poverty and starvation without access to basic necessities.

Meanwhile, the implementation of the SDGs itself is also a challenge (Kumar et al. 2016), among others:

(i) Provides the cost of eradicating global poverty ($ 66 billion), improving infrastructure such as water, agriculture, transportation and energy annually ($ 7 trillion);
(ii) Maintaining peace and stability by combining key factors threatening global stability and security between developed and developing countries;
(iii) Provides a quantitative method for quantifying the target achievement of SDGs that is not yet known and that the degree of measurement depends on the availability of data; and
(iv) Accountability at every level of MDGs input is a deficiency to be aware of when implementing SDGs.

In addition, the implementation of the SDGs also needs to take into account the six transformative challenges of the world TWI2050 (2018) such as:

(i) Strengthening human capabilities and demographics through people-centered development;
(ii) Maintaining a sustainable consumption and production pattern (C&P);
(iii) Decarbonization and energy;
(iv) Enhancing sustainability through food, biosphere and water;
(v) Smart cities; and
(vi) The digital revolution.

By comparison, the implementation of SDGs is more widely covered to the field level compared to the implementation of MDGs for three main factors (Sachs 2012);

(i) Influence of the concept of sustainable development dimensions;
(ii) Increased global awareness as a result of scientific proofs of planetary boundaries; and
(iii) Increasing institutional participation globally.

First, the concept of sustainable development dimensions refers to the process of interdependence between socioeconomic environmental dimensions. SDGs need to work harder to break the "silos" of mono-disciplinary knowledge from every socioeconomic environment aspect through academic support conducting interdisciplinary research in order to come up with ways to measure each achievement and determine new governance approaches that can overcome the achievement of MDGs that are limited to linking social and economic agendas with traditional approaches (Lu et al. 2015; Biermann et al. 2017). The second factor is the increase in global awareness as a result of scientific evidence of planetary boundaries, i.e., the discovery of ozone depletion through chemical studies by Nobel Laureate Paul Crutzen, using the term Anthropocene to refer to current age as the new geological age when human activity is central and threatens the earth's dynamics (Sachs 2012). Thirdly, the increasing participation of institutions globally in reference to a series of different UN conferences and conventions has drawn the attention of important institutions around the world. More than 300 public–private partnerships under the auspices of the UN through the Multilateral Cooperation Initiative were announced at the Sustainable Development Conference (WSSD) in 2002 in Johannesburg (Bäckstrand 2006). Therefore, the brief implementation of the SDGs requires integration of knowledge, global awareness and institutional cooperation to enhance its achievement.

The achievement of the SDGs relies heavily on trade-offs between its seventeen goals by proposing different approaches. In the opinion of Biermann et al. (2017), SDGs are a novel approach because the nature of SDGs' goal setting has never been used before either in the setting of MDGs or in the course of sustainable development. Increased trade-offs of SDGs have also been suggested to adopt a nexus approach that integrates multiple goals into a plan (Weitz et al. 2014; Boas et al. 2016). Then, a cross-sector approach was introduced to replace the single-sector approach adopted during MDGs (Boas et al. 2016; Hazlewood and Bouyé 2018). All of the approaches presented are top-down approaches for the implementation of SDGs nationally and locally.

The top-down implementation of the SDGs is further elaborated by five key steps for increasing trade-offs between SDGs, i.e., first—depending on the number of institutional or sectoral factors that will go beyond the formalization of commitment (making institutional commitment in SDGs more formal), second—strengthening the global order of governance, third—translating the goals of global initiatives into each national context, fourth—integrating sectoral policy and fifth—maintaining the flexibility of governance mechanisms (Biermann et al. 2017). Based on these five steps, it can be concluded that the implementation of the SDGs requires two important mechanisms at two different levels, namely governance at the institutional level and translation mechanism at the local level (Biermann et al. 2017; Hazlewood and Bouyé 2018). Governance mechanism is important for organizing institutions to tailor the role of each institution (important sector) toward achieving the SDGs, while translation mechanism is important to broaden the understanding of sustainable development concept especially to the communities and the public to meet the needs of SDGs.

In addition to focusing on SDGs in terms of their implementation through governance and effective translation, SDGs also place considerable emphasis on civil society participation. The designation of the SDGs as the latest global initiative is by the agreement of 70 representatives of the global civil society compared to the designation of MDGs was only determined by representatives of the UN secretariat (Biermann et al. 2017; Spijkers and Honniball 2014; Sachs 2012). Civil society participation is more widespread when SDGs have a communication channel by NGOs as a representation in dealing with the public (Spitz et al. 2015). It is believed that civil society participation through NGOs in sustainable development has doubled since the establishment of the UN system in the 1950s (Yap 1990).

## 2.2 Characteristics of SDGs

The UN report "The Future We Want" 2012 revealed that the features of the SDGs must be action-oriented, concise and easy to communicate, limited in number, aspirational, global in nature and that their performance is comprehensive across all countries while considering differences in reality, capacities and levels of development and respecting the priorities and policies of a country (Glaser 2012). The SDGs are a new governance approach with seventeen goals inclusive of each other, but the implementation of the SDGs is too loose whereby it is non-legal binding and dependent on weak institutions and made up of leaders of member states of the UN as the enablers that have the freedom in interpreting SDGs according to their respective countries' context (Biermann et al. 2017). The implementation of the SDGs was made more difficult when the eradication of global issues during the implementation of MDGs was not reached beyond 2015, and the unsustainable human dependence on the earth's life-support system has been alarming. Therefore, the definition of sustainable development has been reviewed in accordance with the Anthropocene era which should be "development that meets the needs of the present while safeguarding earth's life-support system, on which the welfare of current and future generations depends" (Griggs et al. 2013). In addition to addressing human welfare dependence on the earth's life-support system, the major difficulty of the SDGs is to resolve the interdisciplinary nature of sustainable development (Glaser 2012). At the same time, not all of the proposed SDGs globally have been used at the national to local level (ibid. 2012). Accordingly, the institutional restructuring process is required to increase participation in the implementation of SDGs.

### 2.3 Institutional Arrangement to Participate in the SDGs

The participation of institutions for sustainable development aims to study the progress of implementing sustainable development policies, to participate in the legal process and to regulate the socioeconomic environmental development at national, regional and international levels (Grybaite and Tvaronavičiene 2008). Meanwhile, increasing participation of social movements from various levels is a sign of increasing global concern about the importance of sustainable development which has led to the process of institutionalization. In this regard, "good governance" is a key recipe that links the impetus for the implementation of goal setting and institutional participation (Biermann et al. 2017; Kemp et al. 2005; Sachs 2012).

The SDGs' reliance on the weak institutional arrangement can be understood as the absence of formalization (no process to make the SDGs more formal) at the institutional level (Young et al. 2017) as the UN 70/1 resolution involved only 193 representatives of UN member states and civil society representatives globally during the UN General Assembly which took place on September 15, 2015, but none at the intergovernmental level. The participation in SDGs from all countries of the world is needed because no country has achieved sustainability despite its advanced country status (Biermann et al. 2017). What is challenging for the implementation of SDGs is the existence of sectoral policies that are not practical at the national level, thus maintaining a single-sector approach during the implementation of MDGs (Glaser 2012), and several previous studies have suggested the transition of a single-sector approach to a cross-sector approach to increase trade-offs between the SDGs (Boas et al. 2016; Hazlewood and Bouyé 2018).

## 3 Non-governmental Organizations (NGOs)

Increasing trade-offs between the SDGs not only requires a shift in sector policy, but also requires a clear implementation mechanism for the SDGs at the institutional and local levels. As a result, the research community has begun to suggest the participation of the civil sector as the primary domain of the SDGs participation in addition to the government, industrial and university sectors for the implementation of the SDGs. The civil sector represented by NGOs, representing civil society groups, is seen as capable of implementing the SDGs at the institutional level as they work to represent a broad range of institutional interests (Doh and Teegen 2002) at the local level as NGOs are able to expand the social aspects of other sectors (Schwartz and Pharr 2003).

### 3.1 Defining NGOs

According to Vakil (1997), the use of the term "non-governmental organization" (NGO) was first introduced by the UN in 1950 and has been defined as self-governing, private and nonprofit organization to improve the standard of living of the less fortunate. The use of the NGO acronym is likely to be used in international relations or the work of developing countries, since its origins date back to the founding of the UN in 1945, when the term NGO was awarded to several non-governmental organizations operating internationally and given consulting status in UN activities (Lewis 2010). The source of this term is derived from Article 71 of the UN Charter of 1947 which declares "non-governmental" that was later introduced in ordinary English language as stated in Resolution 288 (X) of the United Nations Social and Economic Council (ECOSOC) at February 22, 1950, in which the article explicitly referred to the article as the highest legal and legitimate source of justice in the world agreed upon by all member states, while the resolution clearly states that any international organization that is not established under the Treaty of Government must be considered an international non-governmental organization (de Fonseka 1995).

The use of the term "non-governmental" has gained the attention of the academic world in debating the link between the political ideology of anarchism and the emergence of NGO in politics and institutional construction. From the concept of sustainable development, self-governance and voluntary are the most coherent ideas presented by the ideology of libertarian social philosophy as the ideal of political economy to maintain a sustainable resource under some form of governance (Mebratu 1998). It is useful in understanding civil society relations with governments that should not be misinterpreted as a threat by civil society to replace a country's system of governance because civil society only exhibits its own nature through social goals that have existed in the traditional system (Turner 1998). The ideas of the self-governance and voluntary that shaped the concept of sustainable development in terms of political economy were found in the identity of NGOs. It should be noted, however, that NGOs are not based on the ideology of anarchism although the translation of the term NGO is directly using "non-governmental" which is in contrast to "without government" (de Fonseka 1995). Anarchism is a radical ideology aimed to eradicating the bureaucratic system practiced by the public sector, while NGOs are encouraging cooperation among their like-minded counterparts, especially supporting the government's efforts to achieve the aspirations of national development (de Fonseka 1995). In addition, the growth of anarchism in history has been a civil society response to the lack of response in terms of bureaucratic management of the government sector which has a characteristic of its strategic efficiency in administering national resources and social needs (Korten 1984). While the term NGO has a wide range of uses where there are many other overlapping terms used such as "nonprofit," "voluntary" and "civil society" organizations (Lewis 2010), the terms used other than NGOs do not reflect analytical and descriptive rigor, but rather due to cultural and historical differences in thinking NGOs have emerged, such as "nonprofit organization" used more frequently in the USA and the use of "voluntary organization" or "charity" used frequently in the UK (Lewis 2010).

In the post-Rio + 20 era, the UN uses the term "civil society organization" or the CSO acronym in implementing the SDGs. According to UNDP (2014), the civil society must play a critical role in fostering advocacy and transition of policy development, proposing practical solutions and policy opportunities, and criticizing problematic and impractical policies. In the institutional setting of the quadruple helix model, van Waart et al. (2016) also highlighted civil society as the latest domain in policy development and innovation. Therefore, the use of the new term CSO logically covers various levels of nonprofit organizations. In line with the comprehensive implementation of the SDGs that need to integrate civil society from the global level to the grassroots level, according to AlAtas (2003), the civil society is made up of two leading components, the NGO and the non-state actor, whereas the use of the term NGO is one of the organizations' participation in the CSO that has a more organized structure and importance in its development.

## 3.2 The History of NGOs in Development

NGOs have emerged since the eighteenth century in Western countries (Lewis 2010). During the Cold War era from 1947 to 1991, NGOs were seen as part of civil society when balanced relations emerged between governments, markets and the third sector which later became the paradigm of international stakeholders to begin campaigning for the "good governance" agenda (Lewis 2010). This occurs when there is a societal dynamic force that allows many individual opportunities to unite with like-minded individuals (Schwartz and Pharr 2003). Individual opportunities that are combined such as common ideas, needs and causes for promoting a collective gain are known as "collective action" (Olson 1971). When the "collective action" of an individual group continues from time to time in a way that identifies and reflects the importance of social change, it is known as the "social movement" (Teegen et al. 2004). Historically, various social movements, such as religious, cultural and ethnic, have called for the government and business sectors to respond to the wider community, but often the government and business sectors have ignored such interests until too late, and wars and violence have forced them to combine social movements when they are oppressed (Teegen et al. 2004). Communities have lost confidence in institutions in protecting their interests and thus require new mechanisms to step up social capital formation (Putnam 2002). Later, social movements became more organized, influential and integrated into the political system as well as in global business that are often known as new forms of organization, or their interests are translated into structured social movements to form a freestanding organization with other institutions, which is NGO (Teegen et al. 2004).

Since the 1960s, the number and size of NGOs in international development and aid have increased dramatically (Carroll 1992; Clark 1991; Fisher 1997; Fowler 2000; Edwards 1999; Lindenberg 2001). This increase in terms of number and size doubled the role of NGOs in the 1980s until it was celebrated by the international stakeholders as NGOs brought new solutions to long-term problems arising from less effective assistance from government sector in development projects (Lewis 1998, 2010). In other words, the failure of government sector assistance can be termed as sectoral failure or a solution by the government sector having a history of efficiency in its provision of services as a public service product, but ending with market failure and voluntary (Bryson et al. 2006). In addition, sectoral failure can lead to public–private failure when the government determines the price of public goods by lowering the rate of excludability and rivalry of the business sector which can lead to market failure (Brinkerhoff 2003). This has enabled NGOs to improve the public–private sector by playing a role in governance and creating value for social goals (Teegen et al. 2004).

Causes of growth in the number and size of NGOs have little to do with the pressure of international bodies on a country's government to support and include NGOs in national and international politics (Reimann 2006). In fact, one of the reasons why governments and intergovernmental organizations are also promoting NGOs is because NGOs themselves are able to leverage their advocacy and services; as a result today, the world has shifted toward international law enforcement on global issues (Reimann 2006) which is now a global issue where different dimensions and factors are found due to the fact that civil society members are diverse from one another, from individuals to religious and academic institutions to the focus issues like NGOs (Gemmill et al. 2002). "Different" in other words, NGOs have distinctiveness until too many types of NGOs that do not have the same approach in solving problems.

## 3.3 Taxonomy of NGOs

The distinctive nature of NGOs can be illustrated through several NGO evolutionary orientations. Nanthagopan et al. (2016) summarized Korten's (1990) study that NGO evolution has occurred four times. The first generation is oriented to relief and welfare aimed at serving directly in addressing the immediate need of emergency in times of war or disaster. The second generation is oriented to human development that involves the development of the capacities of the local community by fostering self-reliance to meet their needs. The third generation is oriented toward sustainable development to bring about policy and institutional change at every local, national and global level. The latest generation is the fourth generation focuses on people-centered development through social movements and global change.

Human-centered development is the creation of initiative based on human resources that emerged when people-oriented perspective gained widespread attention in the field in the 1970s when developmental performance began to make a direct contribution to improving social and psychological well-being. (Korten 1984). Human-centered development is the main agenda of sustainable development today because the human-oriented perspective was not yet capable in improving the standard of living of human beings in the form of humanitarian efforts, such as poverty eradication, hunger, gender equality, education opportunities and others whereby there are still some gaps between social and economic dimensions (Korten 1984).

The emergence of NGOs in different civil society contexts and the creation of human initiatives from various dimensions and fields have influenced the existence of several NGOs that are trying to improve the system in development. Overall, Teegen et al. (2004) considered all NGOs to be social purpose NGOs because all NGOs initially acted to represent the interests of civil society. In addition to social goals, there are NGOs working at the regional level representing the relationship between developed and developing countries for humanitarian purposes (Lewis 1998) and there are NGOs working to represent social interests within a country (AlAtas 2003). The types of NGOs are summarized in Table 1.

**Table 1** Types of NGOs by context

| Type of NGO | Descriptions | Context | References |
|---|---|---|---|
| Advocacy NGO | NGO with an effective voice in understanding the specific needs of the community, a logical way of consulting community norms in the decision-making process in the event of a conflict between market and ethical needs, shifting a broad and centralized institutional power so that community organizations can participate equally in institutions and providing access to institutions to promote public access to reduce the negative effects arising from the actions of other sectors | General | Teegen et al. (2004) |
| Operation NGO | NGOs with a range of activities involving technical expertise in providing goods and services, expanding the welfare of the community as they have been nurtured to cope with difficult situations with marginalized groups, able to meet the needs of the people when a country is under severe political pressure, indebted or corrupt. Other sectors are difficult to meet the needs of the community, and operation NGOs can determine the scope of operations according to clients' needs whether operating within a country's population or across multiple countries. Such options in determining the scope of operations are not available in the government sector that is bound by the policy of a country even if it operates across the borders of other sectors | General | Teegen et al. (2004) |
| Hybrid NGO | NGOs that have simultaneous advocacy and operation functions or integrated code of conduct to govern public–private activities with the community. NGO's code of conduct can influence the code of conduct of both the government and the private sectors and bargain between both parties | General | Teegen et al. (2004) |
| North NGO (NNGO) | NNGO comes from advanced industrialized countries who take responsibility to provide assistance during emergencies in developing countries | Regional | Lewis (1998) |
| South NGO (SNGO) | SNGO comes from developing countries that receive funding from industrialized countries or receive funding from NNGOs | Regional | Lewis (1998) |
| State NGO | State NGOs are also known as "government-sponsored NGOs" which are managed at the grassroots level by local communities, but at the same time they are under the auspices of the government and bureaucratic governance | Malaysia | AlAtas (2003) |
| Autonomous NGO | Community organizations that are registered either under the registration of companies or the establishment of societies. Autonomous NGOs and sponsored NGOs are different in the early days of their establishment in which they are not by the government | Malaysia | AlAtas (2003) |

## 3.4 Significance of NGOs

Since the 1980s, the growth of NGOs in terms of number, size and taxonomy has gained popularity as "magic bullets" driven from various directions, but has remained development issues as the target (Edwards and Hulme 1995). This popularity has attracted the attention of practitioners from various fields to emphasize the importance of NGOs in development (Lister 2003). This popularity is also linked to criticism by several institutions that doubts on the NNGO's rights as a development player trying to engage in policy formulation and implementation in developing countries (Lewis 1998; Lister 2003).

In the early NNGO humanitarian operations in the late 1990s, their identities were fragile, working in a complex and difficult policy environment as a result of pressure from governments and their supporters in any country they operated. Thus, this skepticism was identified by Lewis (1998) when different levels of organizational autonomy would shape different views between NNGO and SNGO on cooperation for both organizations whereby NNGO wanted to cooperate with SNGO so that is seen as having equal interest at the institutional level. However, SNGO viewed cooperation at that time as an opportunity to access NNGO's resources so that it could rely on resources rather than focusing on sharing development issues together. This skepticism has led to NNGOs perceived as weakening its role as an intermediary rather than providing direct service to the community (Smillie 1994). Lister (2003) described this doubt as the "crisis of legitimacy" experienced by NNGOs. This skepticism has triggered a series of questions on NGOs, such as their role, legitimacy, accountability, representation and performance in explaining NGOs' credibility in development.

Arhin (2016) presented three NGOs' roles in implementing SDGs through the analytical framework based on the research of NGOs' roles by Lewis and Kanji (2009) and Banks and Hulme (2012), including the roles of advocacy, facilitation and brokerage and service provision. Determining the roles of NGOs in implementing SDGs is based on three key issues if NGOs play a critical role in implementing SDGs, namely limited funding, operational disruption and diluted NGOs' identity. Hence, the roles of NGOs in each of the SDGs need to be categorized in line with the priorities and strengths of NGOs so that the implementation of SDGs by NGOs is more focused. For example, social NGOs certainly prioritize the goals of social and human-oriented SDGs, namely SDG 1: end poverty in all its form everywhere; SDG 2: end hunger, achieve food security and improved nutrition and promote sustainable agriculture; SDG 3: ensure healthy lives and promote well-being for all at all ages; SDG 4: ensure inclusive and equitable quality education and promote lifelong learning opportunities for all; SDG 5: achieve gender equality and empower all women and girls; and SDG 6: ensure availability and sustainable management of water and sanitation for all, whereas development NGOs will focus their programs on SDG 7: ensure access to affordable, reliable, sustainable and modern technology for all; SDG 8: promote sustained, inclusive and sustainable economic growth, full and proactive employment and decent work for all; SDG 9: build resilient infrastructure, promote inclusive and sustainable industrialization and foster innovation; SDG 10: reduce inequality within and among countries; SDG 11: make cities and human settlements inclusive, safe, resilient and sustainable; and SDG 12: ensure sustainable consumption and production patterns. Other NGOs such as environmental NGOs certainly prioritize the environmental conservation-oriented SDGs, namely SDG 13: take urgent action to combat climate change and its impacts; SDG 14: conserve and sustainably use the oceans, seas and marine resources for sustainable development; and SDG 15: protect, restore and promote sustainable use of terrestrial ecosystems, sustainably manage forests, combat desertification, and half and reverse land degradation and half biodiversity loss. But all NGOs need to achieve the following SDGs, namely SDG 16: promote peaceful and inclusive societies for sustainable development, provide access to justice for all and build effective, accountable and inclusive institutions at all levels, and SDG 17: strengthen the means of implementation and revitalize the global partnership for sustainable development. Table 2 shows the importance of NGOs in terms of role.

Therefore, focusing on the role of NGOs in achieving the SDGs in accordance with the priorities and strengths of NGOs, then only it is worthwhile to demonstrate their importance in the implementation of SDGs, such as the legitimacy of NGOs. Generally, the legitimacy of an organization increases when its contribution meets the needs of development. The role of NGOs also fills the gaps in developmental stages to enhancing the legitimacy of NGOs as an organization that goes beyond the constituency of other sectors. However, the legitimacy begins to receive attention when the organization undergoes a crisis of legitimacy, as an example, the involvement of NNGO in policy formulation and implementation in developing countries (Sogge et al. 1996). NGOs can enhance the legitimacy of their organizations as long as they are not regulated by membership to make them more responsible, while NGOs need to be more responsible in their implementation if their demands for legitimacy are to be maintained (Edwards and Hulme 1995). Generally, legitimacy is defined as a general perception or assumption that an entity's actions are good and reasonable or in accordance with the norms, values, beliefs and definitions of certain social systems (Suchman 1995). In any

**Table 2** Role of NGOs in development

| The roles of NGOs | Descriptions | References |
|---|---|---|
| Implementer | The role of NGOs that prioritizes the transfer of resources to provide goods and services to people in need | Lewis and Kanji (2009), Banks and Hulme (2012) |
| Catalyst | The role of NGOs as catalyst is defined as the ability of NGOs to inspire, facilitate or contribute to the improvement of mind-sets and actions to promote social change | Lewis and Kanji (2009) |
| Partner | The role of NGOs as partner is defined as a reflection of the development of NGOs working with the government, stakeholders and the private sector in joint activities, such as providing specific inputs on projects or programs of various agencies, or taking on social responsibility in business initiatives | Lewis and Kanji (2009) |
| Advocacy | The role of advocacy is divided into two perspectives:<br>(a) The "Big D" is an NGO advocacy that goes beyond the fundamental changes in the implementation of projects that have a huge impact on the challenges of organizing institutions<br>(b) The "Little D" is a continuous process of advocacy for NGOs, with a radical, systemic alternative by seeking various ways of managing the economy, social relations and politics | Banks and Hulme (2012) |
| Service delivery | The ability of NGOs to innovate and experiment with their prompt services in adopting new programs and most importantly offer the participation and implementation of programs at the grassroots level to foster self-reliance and promote sustainability | Banks and Hulme (2012) |
| Facilitation and brokering | Connecting multiple social, economic and political players in a given task to achieve a goal that is unattainable | Banks and Hulme (2012) |
| The role with government section | Compliance with the law, improving policies that are impractical, advising the need for formal bodies to meet certain external policies, reporting actions to the authorities, becoming a watchdog of any institution, society or individual who tries to act unlawfully | Jepson (2005) |
| The role with industry sector | Promoting and designing corporate social responsibility (CSR) projects, as well as NGO involvement in auditing and monitoring CSR projects, are also a reflection of the changing governance environment for global community domain restructuring as determined by business activities | Arenas et al. (2009) |
| The role with university sector | Emphasizing the potential of civil society that enables the development of new knowledge to address the challenges of a complex world | Maldonado (2010) |
| Globalization agent | NGOs as agents of globalization by:<br>a. Using some of the strategies of international institutions without reservation to globalize their project strategy even though they also criticized the process of globalization (Teegen et al. 2004)<br>b. Exercising the freedom to operate with the advantage of value creation globally when other institutions such as governments and intergovernmental agencies experience paralysis (bound) or limited constituency of administration<br>c. Influencing the involvement of multinational companies in global governance by designing a code of conduct that identifies NGOs as examining and balancing the activities of multinational companies<br>d. Representing the civil society in global governance as the driving force behind international cooperation through the transition of public support toward actively supporting international treaties | Doh and Teegen (2002), Teegen et al. (2004), Gemmil et al. (2002) |

country with social systems in place, NGOs need some technical approaches to improve their legitimacy as shown in Table 3.

Although the role of NGOs is gaining popularity among stakeholders and public trust as a result of overcoming other sector failures and enhancing its legitimacy to meet social goals and market needs, some NGOs also erode institutional and public trusts as a result of a series of public scandals and make excessive claims about their legitimacy as a value-driven organization that they should monitor and evaluate in their achievements (Ebrahim 2003). This excessive claim coincides with the term NGO narcissism or an organized identity that admires its legitimacy in overcoming accountability that such organizations prefer to consider its existence rather than prioritizing its service delivery (Ganesh 2003).

**Table 3** Technical approaches by NGO (Lister 2003)

| Approaches | Descriptions | References |
|---|---|---|
| Legitimacy environment | A heterogeneous environment or an internal and external environment of an organization that has an increase in legitimacy with support and constituency from different partners | Lister (2003), DiMaggio and Powell (1991) |
| Multifaceted nature of legitimacy | The nature of multifaceted legitimacy refers to the various forms of "legitimacy asset" which are divided into four parts:<br>(a) Regulatory—NGOs' legal concerns include compliance with the laws and requirements of the official body which provides sectoral policy through support for the implementation of conventions or formal strategies<br>(b) Pragmatic—the legitimacy of NGOs is based on self-interest calculation of their immediate audience or the feedback of individuals or interested parties that directly deal with NGO activities<br>(c) Cognitive—the legitimacy of NGOs is based on public observation or interpretation toward NGO activities that are based on taken–for-granted status<br>(d). Normative—the legitimacy of NGOs is based on public values and moral standards to assess the role of NGOs that should be in line with public norms | Jepson (2005), Hilhorst (2003), Suchman (1995), Dart (2004), Lister (2003), DiMaggio and Powell (1991), Najam (1996) |
| Legitimacy symbol | A legitimacy symbol that can be identified in an organization and that symbolism can fulfill a partner's value judgment | Lister (2003), Dowling and Pfeffer (1975) |

The importance of NGOs from the standpoint of accountability is emerging to examine how NGOs should treat accountability as a value-driven organization after their actions starting to be judged. Accountability is the process by which an individual or organization reports to the authorities and they need to be responsible for their actions (Edwards and Hulme 1996). Hilhorst (2003) defines NGOs' accountability as a different process player trying to bargain with the benefits and legitimacy of NGOs' activities. From the perspective of dualism, accountability is divided into two, namely internal and external dimensions. Internal dimensional accountability is a sense of responsibility that is manifested through individual actions or organizational mission (Fry 1995). The external dimension of accountability is the responsibility of an individual or an organization's mission to meet a set of standards of behavior (Chisolm 1995). This dualism perspective still lacks in explaining the sense of accountability that focuses solely in compliance with the requirements and formal representation of public institutions, but does not include accountability on behalf of NGOs themselves (Behn 2001; Dunn 1999; Przeworski et al. 1999; Weber 1999). Accountability now exists not only in the relationship of NGOs with the authorities but also in the needs of other sectors such as corporate constituencies. The types of accountability involved in NGOs are shown in Table 4.

The importance of accountability is closely linked to the importance of representation as mentioned by Edwards and Hulme (1996) that NGOs can still be recognized as an organization that is transparent, accountable and acting through the spirit of partnership that has been built with others besides the internal expertise that they work with. This confirms that NGOs can still be responsible representatives without the support of their internal membership as they still have external support for their mission.

Likewise, the implementation of the SDGs needs to look at the importance of NGOs in terms of representation. Representation is defined as a criterion involving NGOs that are often used to criticize both governmental and multilateral development programs (cooperation between government agencies at the regional level) on the basis of procedure, transparency, accountability and participation (Atack 1999). In this context, representation should be viewed as an NGO's interest as an organization representing the interests of civil society. In line with AlAtas's (2003), the civil society is represented by two major groups, NGOs and CSOs. Although NGOs are representative of civil society, there are also theoretical limitations on NGO representation in comparison with governments. Governments reach the whole in terms of both community and spatial space, whereas NGOs respond to specific interests and parts of civil society (Frantz 1987) or meaning NGO constituencies as representatives of civil society according to specific interests and areas of society, while government constituencies include the whole interests and area of society in a state or country.

The performance of NGOs is also important in the implementation of the SDGs in response to their ability to achieve the objectives of the SDGs. In the early 1990s, NGOs' performance was an important factor in development based on three key factors, namely the need for formal assistance by NGOs in influencing the rapid growth of NGO

**Table 4** NGOs' accountability in development

| Types of accountability | Descriptions | References |
|---|---|---|
| Upward accountability | Referring to the relationship with the highest level of management, such as between stakeholders, the founders and the government, and they often focus on allocating expenditure for the intended use | Ebrahim (2003) |
| Downward accountability | Referring to the relationship at the external operation level, such as those who receive direct NGO services, communities or areas that receive an impact by NGO programs | Ebrahim (2003) |
| Internal accountability | Referring to the internal relationship of the NGOs, including the responsibility of the NGO to staff, organizational direction and individuals acting as executors or decision-makers at the field level | Ebrahim (2003) |
| Institutional accountability | Formal accountability that functions to determine the priorities of management and the soundness of the organizational structure | Avina (1993) |
| Heightened accountability | Be one of the components of institutional accountability in formalizing accounting management and organizational audit | Avina (1993) |
| Functional accountability | Accountability exists through aspects of resource use | Avina (1993) |
| Strategic accountability | Accountability exists through the impact of NGO activities | Avina (1993) |
| Structural accountability | Accountability relates to organizational structure | Hilhorst (2003) |
| Public accountability | Accountability relates to public trust | Hilhorst (2003) |
| Rational accountability | Accountability that promotes the practice of transparency of operations is a key focus in the multi-party alignment and interaction | Harsh et al. (2010) |
| Moral accountability | Accountability that promotes evidence of good cause is created behind the scenes so that both can support the development system | Harsh et al. (2010) |

funds, skepticism over the claims of NGOs stating their development programs were more effective than that of government sector, and the shifting role of NGOs that led to the increasing demands of organizations in line with their achievements in order to redefine their role (Fowler 1996).

The performance of NGOs is also influenced by the duration of funding from government subsidies or corporate donations as there is evidence of poor NGO performance due to short-term financing, but there are also NGOs depending on the level of democracy which is free from external interests, near to the poor and willingness to face anyone in power (Edwards and Hulme 1995). Today, NGOs are also impacting the performance of other stakeholders, especially involving law enforcement by the government through the provision of NGO reporting, which benefits the corporate sector in increasing their profits through corporate tax deductions, and the excess paid tax can promote corporate voluntary activities and indirectly reduce the cost of corporate operation (Zainon et al. 2014).

However, the performance of NGOs also needs to be seen in terms of the type of institutional network that cooperates with them as most NGOs have informal networks where informality is a problem when there is an increase in the size of NGOs' networks with their stakeholders and the community (Atack 1999). In implementing the SDGs, NGOs also have to play a role in enhancing the formalization of commitment at the institutional level and need to reduce informal interruptions. Most NGOs operate in horizontal organization form and informal manner, while the government sector practices, and hierarchical and vertical organizational forms or operations in government sector need to follow the levels of authority and executive of an individual in the field of management (Gordenker and Weiss 1995).

## 4 Concepts

In previous studies, the importance of NGOs in the implementation of the SDGs focused only on the role of NGOs, while the importance of NGOs in terms of legitimacy, accountability, representation and performance of NGOs was limited. Furthermore, the roles of NGOs presented by Arhin (2016) have not been sufficient in assisting NGOs toward achieving the SDGs, but only inform the limitations of NGOs faced in implementing the SDGs. Therefore, the role of NGOs and the legitimacy of NGOs are the key components of NGOs' organizational capabilities. Meanwhile, the other importance of NGOs, such as accountability, representation and performance, is only supporting components. In addition, the process of matching SDGs' toward their target needs is difficult to determine NGOs' accountability, representation and performance, either quantitatively or qualitatively, while accountability and representativeness are partly justified by Jepson (2005); however, NGOs' performance is not the priority just yet because the SDGs in the first phase of the five-year implementation period (2015–

2019) are more focused on determining the role and legitimacy of NGOs in line with the SDGs' target needs. Studies on NGO performance can be made if SDGs have monitoring mechanisms for measuring organizational performance, but there is still no monitoring mechanism developed by the UN for the implementation of the SDGs.

As for the SDGs implementation process, previous studies only gave the idea of implementing the SDGs require strengthening of institutional capacity (Hezri 2016); nexus approach (Boas et al. 2016); cross-sector approach (Hazlewood and Bouyé 2018); and formalization of the SDGs commitment at the institutional level (Biermann et al. 2017). However, these approaches are not detailed on how institutionalization can be implemented, and no organizational and institutional theories can be applied to understand the requirements of the SDGs implementation process. There are two processes for implementing the SDGs, namely good governance (Sachs 2012) and the SDGs translation. Therefore, the process of implementing the SDGs requires two mechanisms, namely governance and translation mechanisms according to the needs of the SDGs at the institutional and local levels. Cross-sector partnership is proposed as a governance mechanism at the institutional level as we understand that good governance cannot be defined by its type of governance because it does not have the same governance practices at every level of institution (Sachs 2012). On the other hand, broadening social value is proposed as the SDGs translation mechanism which is suitable for different types of social interactions and requires translation according to the level of understanding and acceptance of the local community regarding the implementation of the SDGs.

## 4.1 Organizational Capacity

The scope of NGOs' organizational capacities comprises two key components, namely the roles and legitimacy assets of NGOs (Fig. 1), because these two components are closely related to the internal affairs of NGOs compared to accountability and representation which are more appropriate at the institutional level, while performance is the impact component of the NGOs' program which is within the scope of sustainable development dimensions.

NGOs' organizational capacities refer to the internal level of NGOs in demonstrating as a third sector, functioning as an organization with its own characteristics. Fowler (1996) defined NGOs' organizational capacities as a measurement of the ability of NGOs to satisfy and influence their stakeholders. This definition was proposed when Fowler (1996) initially defined organizational capacity as an organization's ability to effectively achieve a goal in what it set out for implementation and asserted that organizational capacity is not something that can be observed internally because organizational ability gives effects at the external level. Therefore, evaluations need to be made at the external level because NGOs need to follow the expectations of the relevant stakeholders working with them. In this context, organizational ability should refer to the needs of the SDGs as an external NGOs' interest. The need for the SDGs is to formalize the commitment of governance at the institutional level to increase trade-offs between the SDGs (Biermann et al. 2017).

## 4.2 Institutional Capacity

The scope of NGOs' institutional capacities comprises three key components, namely cross-sector partnership, broadening social value and the SDGs participation domain (Fig. 2). Cross-sector partnership component is placed in the institutional domain as the mechanism of governance of the SDGs at the institutional level, while the broadening social value component is placed in the social domain as the mechanism of translating the SDGs at the local and public levels. Both mechanisms are organized by domain to suit the needs of the SDGs at the institutional and local levels. The components of the SDGs participation domain are based on the quadruple helix model which includes the participation of all four institutions in the implementation of the SDGs,

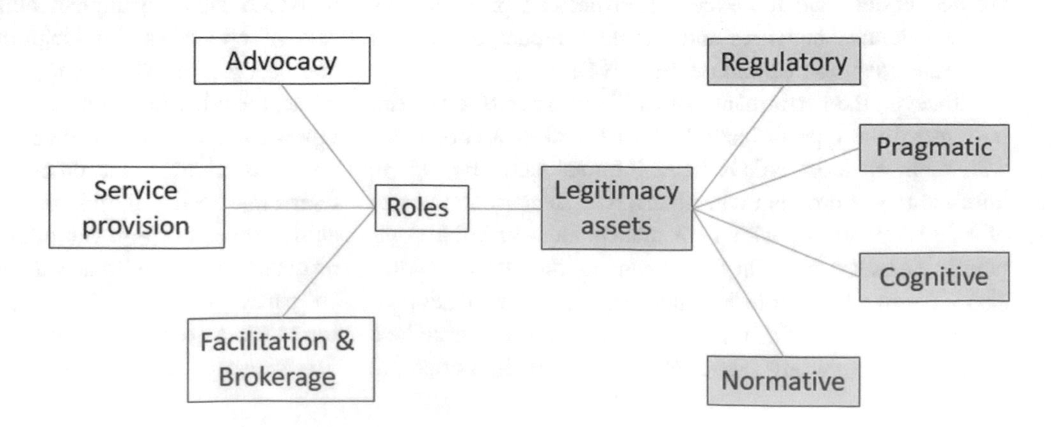


**Fig. 1** Scope of NGOs' organizational capacities (Arhin 2016; Jepson 2005)

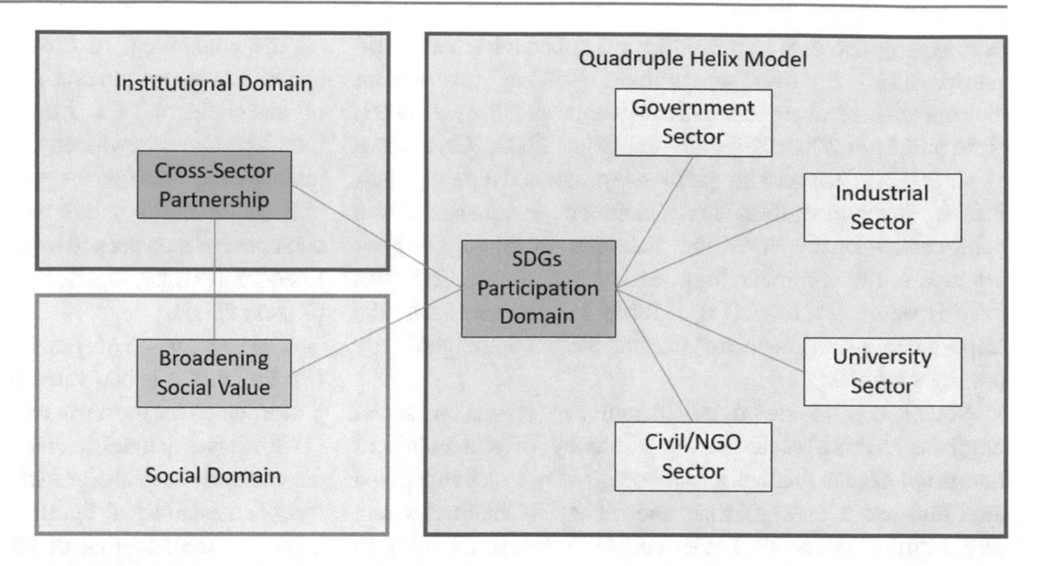


**Fig. 2** Scope of NGOs' institutional capacities (cross-sector partnership mechanism (Googins and Rochlin 2000); broadening social value mechanism (Teegen et al. (2004); SDGs participation domain (Maldonado (2010))

comprising the government, industry, university and civil sectors.

The scope of institutional capabilities is in line with institutional capacity concept by Healey (1998) which exhibits three-dimensional relationships, namely intellectual capital (IC)—knowledge source (K), social capital—relational resource (R) and political capital—mobilization capabilities (M) for institutional capital formation through the formulation of public policy strategies and practices (Fig. 3). Institutional capacity (IC) does not have a definite definition, but Healey (1998) has defined five indicators of institutional capacity building on the concept of institutional capital formation:

(i) Integration of various economic, social and environmental agendas;
(ii) Policy-making collaboration;
(iii) Wide involvement of various organizations of interest;
(iv) Appreciation of various forms of local knowledge; and
(v) Construction of related resources.

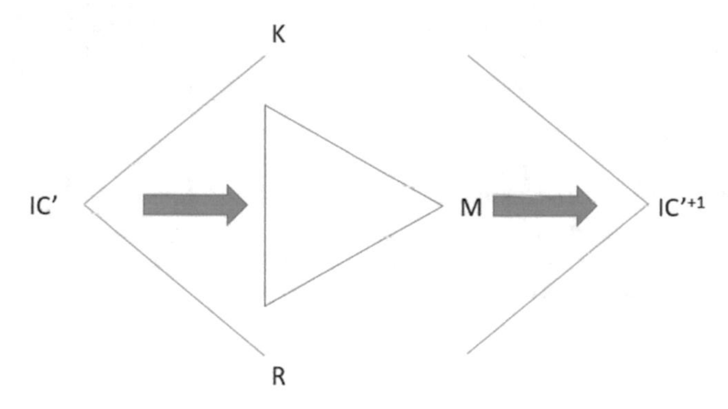


**Fig. 3** Institutional capacity conceptual framework (Healey 1998)

The importance of the involvement of institutional capacity in the implementation of the SDGs is necessary because it is feared that there is institutional isomorphism or imitation of inter-institutional roles that may lead to duplication and overlapping of efforts (Boyer 2000; DiMaggio and Powell 2004; Espey et al. 2015) and may lower trade-offs between the SDGs. Institutional isomorphism is closely linked to the behavior of institutions that try to complement their institutional legitimacy so that it is seen through the eyes of other stakeholders, but in fact it diminishes the value of creativity and innovation practice of institution (DiMaggio and Powell 2004). Thus, the three dimensions of Healey's (1998) institutional capacity concept can be linked to each component of NGO programs, namely cross-sector partnership, broadening social value and the SDGs participation domain.

Intellectual capital—knowledge source (K)—is a platform for the learning environment of the relevant organizations to gain experience in different sectoral relationships in order to form a collaborative approach. The cross-sector partnership component can serve as a continuation of the collaborative approach as proposed by Healey (1998) in which the cross-sector partnership component is an intellectual capital which is also an institution-wide communication platform for the integration of knowledge from each institutional partner toward achieving the SDGs.

Previous scholars' views on cross-sector partnership were different, whereby it is a new intermediary to balance the roles and responsibilities that each community institution plays and to innovate various institutional interests (Googins and Rochlin 2000) into a complex and challenging public problem-solving strategy (Agranoff and McGuire 1998; Goldsmith and Eggers 2005; Kickert et al. 1997; Mandell 2001; Rethemeyer 2005) and methods of dealing with serious social problems (Bryson et al. 2006). Cross-sector partnership also acts as a multilateral collective (multi-party participation) involved in problem solving, information sharing and resource allocation (Kenis and Provan 2009; Rein and Stott 2009; Seitanidi and Crane 2009; Koschmann et al. 2012). Cross-sector partnership can serve as a pluralistic approach to sustainable development governance for the achievement of the SDGs that links the complexity of governance in the administration, market and civil society (van Zeijl-Rozema et al. 2008). Table 5 lists some of the importance of cross-sector partnership as highlighted by previous studies:

Social capital—relational resources (R)—is a social platform that refers to the constituency of a network of interested bodies that has a wide range of network in a given area and has a level of trust and ability to interpret social world differences around other constituencies of the relevant organization (Putnam 2002). Sørensen and Torfing (2007) summarize social capital as building trust in social interaction in civil society.

The component of broadening social value as a social capital has a continuation in the constituency of the network of stakeholders as a form of social value called social interdependence, whereby it is derived from collective interests through increasing public trust. Public trust is the observation of the public based on the quality and value that characterize a social movement, such as the right of speech, freedom, honesty, idealism, cost-effectiveness and efficiency (Jepson 2005).

The importance of social value has increased over the last two decades as social value has become the measurement of performance for most organizations. McClintock and Allison (1989) have defined social value as the essence of social dependency in decision making. Social value measurement has been established by all major institutions of the society through demonstration of NGO project impacts, especially broadening social value as a component of project impact measurement that can be seen by the authorities, corporations, like-minded partners and communities involved.

**Table 5** Importance of cross-sector partnership in development

| Importance of cross-sector partnership | Description | Sources |
|---|---|---|
| Level of collaboration | Cross-sector partnership has three stages to showcase the maturity of a partnership:<br>(i) First stage—reciprocal exchange<br>(ii) Second stage—development value creation<br>(iii) Third stage—symbiotic value creation | Googins and Rochlin (2000) |
| Design and implementation of cross-sector partnership | A cross-sector partnership framework consisting of five components:<br>(i) The initial stage of collaboration—the general environment, the failure of the sector and the direct antecedents<br>(ii) The process of working together—formal and informal<br>(iii) Structure and governance—formal and informal<br>(iv) Contingency and common constraints<br>(v) Outcomes and accountability of collaboration | Bryson et al. (2006) |
| Collaboration of various organizations of interest in sustainable development | Sustainable development requires the organization of institutions through collaboration of various stakeholders in:<br>(i) Different forms of collaboration—transnational, multinational, multi-sectoral and public–private<br>(ii) Collaboration legitimacy—the legitimacy of procurement (input legitimacy) refers to strengthening of institutional capacity and accountability, and the legitimacy of output (output legitimacy) refers to the level of recognition of several network institutions and memorandum of agreement | Bäckstrand (2006) |
| Understanding cross-partnership constituencies | A conceptual framework of cross-sector partnership that forms constituency from the point of view of the communication process and the explanation of the enhancement of cross-sector partnership values through communication practices:<br>i. Conversation—observable interaction<br>ii. Text—communicates symbolically or metaphorically<br>iii. Orientation—the circulation between conversation and text | Koschmann et al. (2012) |
| Cross-sector partnership to address social issues | Social issues swirled around four sectoral arenas; business–nonprofit organization, business–government, government–nonprofit organization and trisectors. The discussion of these four arenas is based on three main platforms for establishing collaboration:<br>(i) Resource-dependent platform<br>(ii) Social issue platform<br>(iii) Society sector platform | Selsky and Parker (2005) |

The constituency of the network of interested organizations or social dependency greatly influences the legitimacy of NGOs by establishing public trust as a legitimacy environment by forming social recognition of the roles and actions of NGOs. On the other hand, it can happen if the public's confidence in the roles and actions of NGOs does not reflect the value of their responsibilities because trust and accountability are closely linked to one another and can undermine the legitimacy of NGOs (Jepson 2005).

In the early 1980s, public trust opened the way for the promotion of various "social interests" in the development agenda through the need for consultation and participation with the public (Healey 1998). According to Adlerian theory, social interest is defined as a sense of community, an orientation toward living with others and a lifestyle that values the common good over one's own desires and interests (Adler 1970). Social interests can create contradictions or conflicts when they have fixed positions in influencing public belief, and in turn social interests can also be the basis for enabling collective actions to generate social values (Healey 1998). Messick and McClintock (1968) outlined three social value orientations, namely cooperation, individualism and competition as a general need of individuals in making decisions that affect not only the personal interests of the people but the interests of those around them (social interests).

Social values through social movements can also be interpreted as social dependency, whereby it is referred to as collective actions through the participation of individuals who have a social value orientation and who are trying to achieve a goal that cannot be achieved on their own (Teegen et al. 2004). Consequently, social movements are formed by the collective actions of a group of individuals who can be identified for a period of time until their actions begin to reflect the importance of social change (Teegen et al. 2004). The role of social movements is often in line with the institutional environment (Sjöstrand 1992). A new model of socioeconomic development is emerging around the role of the institutional environment in which the private, government and civil sectors play a key role in shaping sustainable communities (Googins and Rochlin 2000).

The broadening social value component is adapted from Teegen et al. (2004) idea of the role of NGOs in value creation (not specific to social values alone). Teegen et al. (2004) provided several examples of values created by collective actions by NGOs, such as sustainable development initiatives, global concern, human rights, trade dispute resolution, social welfare along with economic value creation, globalization, the efficiency of a firm's market operations and global equity development for income generation in poor countries.

In making broadening social value in the SDGs' implementation mechanism, social value is placed as the key ingredient in the creation of a centralized human development, namely the human capacity is a central focus of development to improve community self-esteem (self-reliance or reduction of dependence on aid), social justice, participation in making decisions (Korten 1984) on two new educational subjects, namely global citizenship and environmental citizenship.

Global citizenship is a dynamic expression of economic, cultural and ecological integration that brings the human experience to the forefront of the modernization phase of civil society relations (Falk 1993). Global citizenship is more than a process of learning about complex global issues, such as sustainable development, conflict and international trade interest—these are all global dimensions of local issues, as they occur in our lives, areas and communities (Bojang 2001). Three components of the national citizenship education curriculum are outlined by Oxfam (1997), the first component—knowledge and understanding, which is mastering concept, the second component—skills in critical thinking, argument, resolution and challenging skills, and the third component—has values and attitudes from angle of commitment, respect, attention, sensitivity and self-esteem.

Environmental citizenship, on the other hand, sees positive change from the individual level to the collective behavior of the community and institution. Environmental citizenship is not like a fiscal self-interest approach; it is a model of human motivation when society contributes something to their own interests whether in the form of rewards or virtual security embedded in environmental policy by making self-interest a driver of behavior as if to promote environmentally sound behavior (Dobson 2007). On the other hand, environmental citizenship should be a positive change beyond the self-interest approach as it openly ignores public good sustenance such as the environment (the environment as a major provider of natural resources for human life and social and economic purposes) (Dobson 2007). Therefore, broadening social value is a mechanism for the implementation of the SDGs that can anticipate human ability to centralize behaviors and attitudes toward the formation of a better collective commitment, while this mechanism can be applied by NGOs using their capacities at the organizational and institutional levels to influence stakeholders and the local communities to implement the SDG for ecological footprint reduction.

Political capital—mobilization (M)—refers to an individual's power to act politically through participation in an interactive political process (Sørensen and Torfing 2007). Linking the SDGs' participation with political capital refers to the autonomy of sector leaders and key institutional players in establishing institutional capital to use politics as an important step toward implementing the SDGs at every level of the society.

The main challenge of the SDGs at the institutional level is that the political economy of the past still has an influence on the current institutional relationship for "new institutionalism" from the various angles of governance and political ideologies making it difficult for the ruling class (government and administration) to try to control community political strategies in a particular area (Healey 1998). However, this is not fixed and difficult as the local political community is now more dynamic, motivated, adaptable and changeable in its manifestation or interpretation of social relations (ibid. 1998) as they shape agencies' behavior and vice versa (Anthony Giddens 1984).

The most important and challenging part of building the political capital for implementing the SDGs is determining the autonomy of each sector leader and the key players in integrating their efforts to meet the six transformative challenges of the SDGs. Thus, new institutionalism emphasizes the process of institutional change through institutional reorganization and it does not reject the importance of explaining the need for institutional change from the standpoint of classical institutionalism as the SDGs have become a global agenda in need of institutional shifts toward making a more prosperous world change in addition to the need for the SDGs to have a clear mission.

Several years ago, research has shown the importance of participation of key community institutions in the implementation of the MDGs. Arranging institutions involving the three major sectors of society has been manifested through the linear model of the triple helix model, which are the academic, business and public sectors (Etzkowitz 2008) (Fig. 4).

Due to the increasing complexity of development challenges in addressing the capabilities of each sector (Kolk et al. 2008), Maldonado et al. (2009) proposed the integration of the civil sector into institutional mechanisms where civil society demonstrates their potential to generate new knowledge in development to address problems in an increasingly challenging world. The civil sector is represented by two components, namely NGOs and civil society (AlAtas 2003). The importance of NGOs to represent civil society is then manifested in the setting up of a new institution known as the fourth circular (quadruple circle) model that demonstrates civil society participation as one of the most recent domains of the SDGs (Maldonado 2010) as shown in Fig. 5.

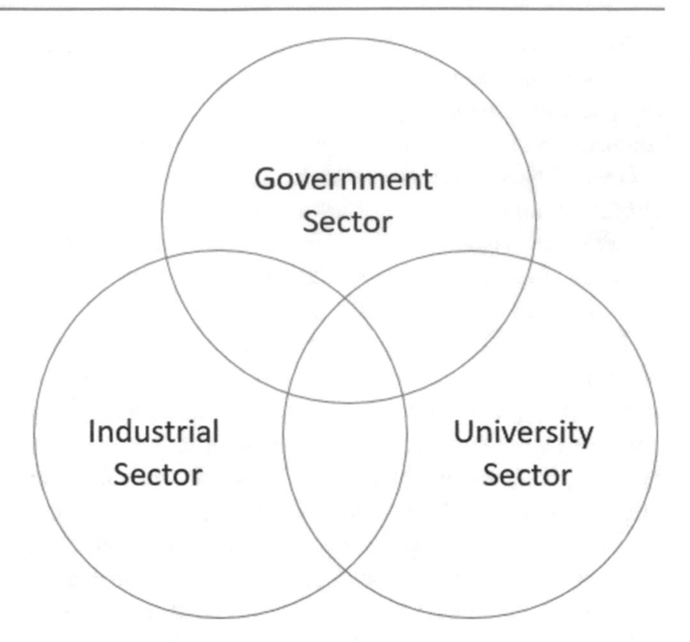


**Fig. 4** Triple circle model for science innovation policy Etzkowitz (2008)

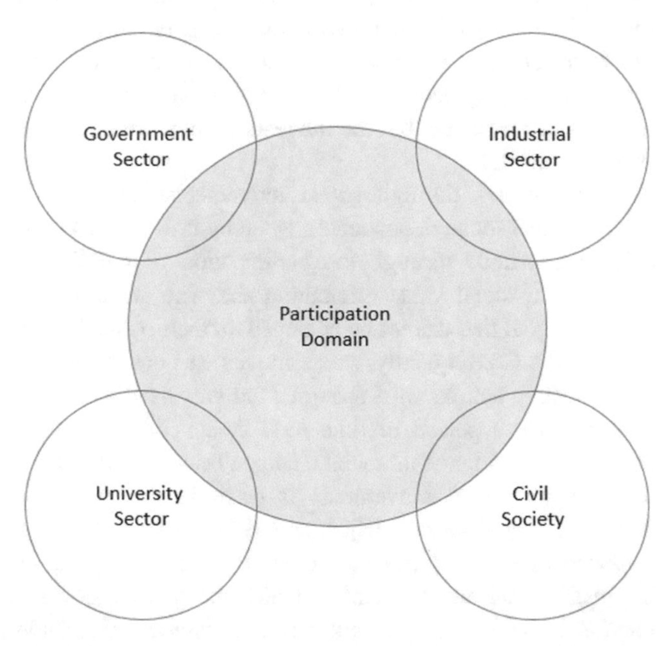


**Fig. 5** Quadruple helix model for science innovation policy (Van Waart et al. 2016)

## 4.3 Sustainable Development

The last component is the impact of the NGOs' programs where the streamlining of the NGOs' programs begins with its organizational capacity, followed by the institutional

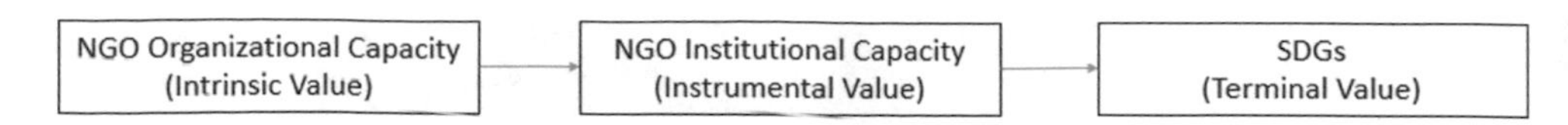


**Fig. 6** Links between the values

capacity to have value and impact on development. In terms of value, there are three values, namely intrinsic value for NGO capacity, instrumental value for NGO institutional capacity and terminal value for sustainable development dimensions (Fig. 6).

The intrinsic value of NGO's organizational capacity is better defined as "in itself" than "for its own sake" as Beardsley (1965) emphasized the tendency to use intrinsic value more effectively to illustrate the meaning "by itself" referring to a potential or advantage that exists in an object or other. The term "magic bullet" by Edwards and Hulme (1996) is referred to as intrinsic value that can be described as the potential or advantage of an NGO in terms of role, performance and accountability if shot from any direction, intrinsic value of NGO still relevant to each direction of development. The implementation of the SDGs in promoting the role of NGO as a representative of civil society is also based on the intrinsic value of the NGOs as the practice of the third sector does not expect for return.

The instrumental value of NGO's institutional capacity has to do with the scholarly discussion of institutional economics by Thorstein Veblen on social value as instrumental value. Instrumental value is a value inherited in a process of connecting other complex values that result from various forms of social interaction to achieve a goal (Tool 1993). The complex value formed by social interaction is called social value, and the criteria for determining social value are valuable and useful to prominently promote their use throughout the concepts, theories and models that guide institutional arrangement processes (Ramstad 1989). To the extent that NGOs rely on a variety of organizational interests (Doh and Teegen 2002) and dominate the social aspect over other sectors (Schwartz and Pharr 2003), social value should serve as instrumental value played by NGO institutional capacity to connect the intrinsic value of NGO capability toward achieving the SDGs as a terminal value.

The SDGs are a terminal value because the SDGs are made to be the ultimate goal by focusing on the organizational and institutional goals to make a significant impact in achieving sustainability. Although the process toward achieving sustainability is never-ending, it is still considered as a paradox and continuous in nature whereby sustainability is an ideal concept today to enhance the knowledge of the global community in dealing with anthropogenic effects from various angles and corners of the globe. The terminal value of the SDGs can be divided into three dimensions of sustainable development, namely social inclusion, economic growth and environmental protection as major pillars of sustainable development.

Since the sixteenth century, gross domestic product (GDP) has been a measurement of economic growth and development, while today's GDP concept is the result of American economist Simon Kuznets. However, modern GDP still lacks in measuring a country's effectiveness in addressing environmental and social issues as modern GDP only emphasizes economic growth in terms of mass production from a country. Today, the Social Progress Index 2015 presents several dimensions of economic performance complexity and social progress that do not exist in the modern GDP (Stiglitz et al. 2017). From a social point of view, social inclusion is the ability of the community to meet their basic needs, establishing a building block that enables the community to enhance and sustain their lives and create the full potential for all (Imperative 2015). From the economic point of view, economic growth is synonymous with modern GDP which is the increase in the amount of goods and services produced by each population over a period of time. Understanding economic growth based on a quantitative approach to sustainability can lead to a loss of focus on social equality (Hezri and Ghazali 2011). Thus, green economy is presented as a new form of economic transition taking into account environmental risks and scarcity of natural resources and biological diversity (Hezri and Ghazali 2011).

All measurements, such as Social Development Index 2015 and the transition to green economy, are aimed to integrating the three dimensions of holistic sustainable development as defined by Brundtland Report on the three-dimensional relationship of sustainable development that depends on ecological balance by taking into account biosphere capacity attempting to absorb anthropogenic effects, reflecting economic growth based on equitable sharing of resources with the poor and social inclusion through human ability to uphold right and justice to claim something that does not benefit the economy and social equality in human development (Robert et al. 2005). A detailed description of each scope of NGO capacity, NGO institutional capacity and sustainable development dimensions along with their respective components and related theories is formulated in a conceptual framework as shown in Fig. 7.

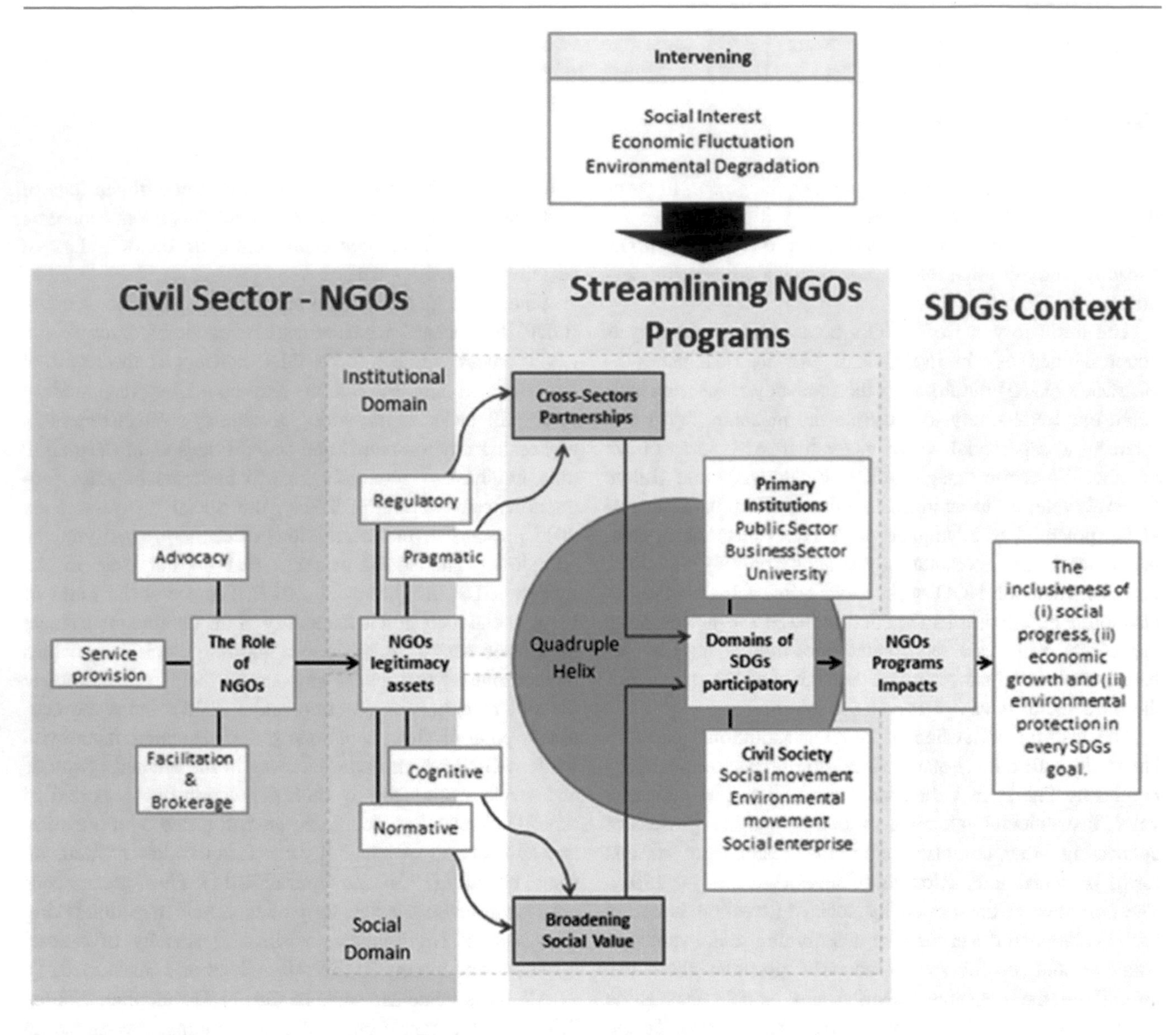


**Fig. 7** Conceptual framework of streamlining NGO's program toward achieving the SDGs (Hassan et al. 2019)

## 5 Case Study: Global Environmental Centre

Malaysia is a country of 30 million populations consisting of diverse races, languages and religions, enjoying significant economic and social progress among Southeast Asian countries (Jayasooria 2016). The role of community organizations in the background of ethnic diversity in Malaysia has been rooted before the term NGO was introduced. Even there are laws during the colonial era to restrict some social movements against the government then; until the present day, it is evident that tensions between the civil societies and the government still exist (Weiss and Hassan 2003).

In modern era, tensions arise from the ideological antagonism that exists between governments and NGOs, just as NGOs find it difficult to cooperate with governments when they are involved in biased and corrupt practices. Whereas, NGOs have been perceived as anti-government by governments (AlAtas 2003). This tension also has to do with the political system in Malaysia, for example, freedom of speech and assembly and other antidemocratic indicators showing that the Malaysian government is practicing "quasi-democratic" that it rests between democracy and authoritarianism in controlling civil society behavior causing this tension to occur on the basis of political interests (Hooi 2013). Therefore, the emergences of NGOs in Malaysia are of great importance as the background and movement of NGOs in Malaysia are unique and different in terms of their origin when compared to Western countries. As for the implementation of the SDGs, even though civil society in

Malaysia is receptive and the Malaysian government is ready to provide the framework for implementing the SDGs (Jayasooria 2016), NGOs and governments need to step in and cooperate to help Malaysia in shifting the single-sector approach to the cross-sector approach as a mechanism of institutional governance to achieve the objectives of the SDGs.

In shaping the relationship between NGOs and governments, disclosure of information by NGOs is a fundamental requirement for accountability (Zainon et al. 2014), especially the SDGs accountability framework to be met by the Malaysian government to the UN delegation to provide measurement and data availability on each SDGs target achievement (Ocampo 2015). Generally, NGOs in Malaysia operate locally, nationally or internationally and the existence of their entities is significant in this country, especially in terms of disclosures related to transparency and accountability from other stakeholders (Zainon et al. 2014). In addition, the role of NGOs in information disclosure also has an impact on Malaysia's foreign policy (AlAtas 2003).

By law, the Societies Act 1966 and Societies Regulations 1984 are the laws governing the conduct of NGOs in Malaysia (Zainon et al. 2014). Some NGOs have registered under certain acts like Universities and University Colleges Act 1971, while others have registered themselves under the Companies Act 1965 (Zainon et al. 2014). Legal information is crucial for the SDGs to review the effectiveness of Malaysia's institutional policies and systems as well as to aim at encouraging NGOs toward legislative compliance or facilitating regulatory processes by the agencies involved.

According to Soh and Tumin (2017), nearly 42% increase in the number of NGOs was recorded from 2002 to 2017 and the state with the highest number of NGOs in Malaysia was Selangor (Table 6). The increase in the number of NGOs and the distribution of NGOs in several economic and municipal areas, such as Selangor, Federal Territory, Johor, Perak and Penang, has subsequently revealed that these states have more than 4000 NGOs, respectively.

The Global Environment Centre (GEC) is an international NGO headquartered in Petaling Jaya, Selangor, Malaysia. GEC is an environmental conservation-oriented NGO, whereby they make environmental issues as one of the most important issues in the world. The establishment of GEC on December 6, 1998, can be classified as a middle-class NGO born when Malaysia became a rapidly developing country in the Asian region. Unlike other NGOs, such as the Environmental Protection Society Malaysia (EPSM), the *Sahabat Alam Malaysia* (SAM) and the Malaysian Nature Society (MNS), they are early pioneers of the environmentalists who have been around earlier than GEC and more aggressive in defending the ideology of environmentalism. There are also relatively new NGOs like EcoKnights and *Pertubuhan Pelindung Khazanah Alam Malaysia* (PEKA), and their

**Table 6** Total NGOs registered in Malaysia (Soh and Tumin 2017)

| States | NGOs registered |
|---|---|
| Johor | 7591 |
| Kedah | 2764 |
| Kelantan | 1311 |
| Melaka | 1992 |
| Negeri Sembilan | 2320 |
| Pahang | 2433 |
| Penang | 4517 |
| Perak | 5722 |
| Perlis | 358 |
| Selangor | 11,878 |
| Terengganu | 929 |
| Sabah | 3223 |
| Sarawak | 3870 |
| Wilayah Persekutuan | 8660 |
| Total | 57,568 |

establishment is based on the development of new opportunities and current environmental issues in Malaysia that are more specific. The like-minded approach is the key GEC approach in establishing partnerships with several other important institutions. In contrast to NGOs adopting ideological contradictions, in which provoking as one form of advocacy, GEC is more open by providing a platform between NGO partners, local communities and the public to collaborate with important institutions so that there is no obstacle in lifting the interests of organizations and local people who were directly involved in their projects after learning much about the adverse effects of provocative actions by other NGOs.

GEC operates in 15 countries, and most of the GEC projects are concentrated in Malaysia and Southeast Asia. As a 20-year-old middle-aged NGO, GEC's operation is more of South-south cooperation-oriented as they work to support the process of exchanges of resources, technology and knowledge between developing countries in the Southeast Asia. In the Southeast Asia, tropical peat conservation is one of the GEC's locally recognized expertise by international bodies, such as the European Union (EU), especially in the area of interest of neighboring Malaysia. On the issue of forest fires contributing to the formation of haze across the country, exploitation of land for oil palm plantation have detrimental effects on the ecological and hydrological values of tropical peatlands in two of the largest oil palm-producing countries, Malaysia and Indonesia. At the global level, GEC is leading a project on degraded peat conservation to stimulate the global community's debate on the applicability of peat conservation policies at each international convention, such as the Convention on Biological Diversity (CBD), the

United Nations Framework Convention on Climate Change (UNFCCC) and the United Nations Convention to Combat Desertification (UNCCD) that can design work and create action plans (Parish et al. 2008).

## 6 Strategies for Mainstreaming, Institutionalizing and Translating Sustainable Development Goals

There are three strategies proposed, namely mainstreaming the SDGs in the goals of GEC and partner institutions, institutionalizing the SDGs in the capacity of GEC institutions and translating SDGs into every dimension of sustainable development. Table 7 is a mapping of the strategies formulated, the conceptual components and the impact of GEC program needed to achieve the SDGs. The program impact for the strategy mainstreaming the SDGs in the goals of GEC and partner institutions is action-oriented because GEC program orientation, direct and indirect roles, field of work of GEC institution partners in accordance with the SDGs and sector role can be translated into action as GEC and its institutional partners have integrated plans in the form of integrated management plan (IMP) or action plan (AP). Therefore, the SDGs need to be included in a more comprehensive integrated plan presented by a number of key NGOs in Malaysia and a number of agencies related to implementing national strategies to achieve global interests. For example, GEC has developed an integrated management plan for North Selangor Peat Forests Volume 1 and 2 to manage peat forests in the state of Selangor with the Selangor State Forestry Department and GEC used the IMP as a strategy for preparing papers, planning plans, action plans and project budget calculations. The IMP also outlines some of the global interests that need to be achieved, such as Aichi Biodiversity Targets and Sustainable Forest

**Table 7** Mapping of concept components, strategies and program impacts of GEC

| Aspects | Concept components | Strategies | Program impacts |
|---|---|---|---|
| GEC program orientation | NGO program impact | Mainstreaming the SDGs in the goals of GEC and partner institutions | Action-oriented |
| GEC direct role | NGO roles | | |
| GEC indirect role | | | |
| Field of work of GEC institution partner in accordance with the SDGs | SDGs participation domain | | |
| Sector role | | | |
| Legitimacy environment of GEC | Legitimacy of NGO | Institutionalizing the SDGs in the capacity of GEC institutions | Formalization of commitment |
| Legitimacy asset of GEC | | | |
| Legitimacy symbol of GEC | | | |
| Workload of organization | Cross-sector partnership | | |
| Institutional capacity gap | | | |
| Form of cooperation, level of cooperation and benefit of corporation | | | |
| Organizational openness and sector selection appropriate for informing and implementing SDGs | SDGs participation domain | | |
| Sector issue | | | |
| Knowledge of SDGs: sources of knowledge and ambiguities related to SDGs | Cross-sector partnership | Translating SDGs into every dimension of sustainable development | Environmental citizenship |
| Readiness of SDGs: SDGs announcement, internal and external promotions of SDGs | | | |
| Context of broadening social value | Broadening social value | | |
| Sector requirements | SDGs participation domain | | |

Management (SFM). If an integrated SDGs plan can be produced, then all the papers, implementation plans, program planning and monitoring activities can be supervised with only one reference being the integrated SDGs plan.

The program impact for the strategy institutionalizing the SDGs in the capacity of GEC institutions is the formalization of commitment or the establishment of institutional networks linking institutional commitments as the legitimacy of NGO, cross-sector partnership and SDGs participation domain reflect GEC's relationship with other sectors and suggesting hybrid governance as a form of SDGs governance within the capacities of the GEC institutions as most of the institutional partners of GEC have a form of public–private cooperation. Public–private partnerships are largely influenced by government actions that have begun to increase the number of private sectors in the provision of public services through privatization, service contracts and social infrastructure allocation (Li and Akintoye 2003). Therefore, formalization of commitments among GEC institutional partners for the implementation of the SDGs should take into account the role of other private sectors which have the linkages with governmental sectors in terms of public service delivery function as many GEC institutional partners choose reciprocal exchange cooperation in which forming cooperation between institutions is only for the interest of their respective organizations. The integration of diverse policies as well as resources and skills pools demonstrates public–private cooperation in the capacities of GEC institutions to rely on public service demands. The reliance on resources, such as policy compliance, funding and resource support, is a form of GEC dependence with other sectors. Therefore, cross-sector partnership governance mechanism in the capacities of the GEC institutions is a public–private partnership that relies on the fulfillment of public service demands in terms of policy compliance, funding and resource support.

The program impact for the strategy and translating SDGs into every dimension of sustainable development is a social value that is essential in fostering environmental citizenship. Environmental citizenship can be considered as one of the components of global stewardship in meeting SDG 4 requirements, education priorities at every social level. The subject of global citizenship not only needs to be emphasized solely in pedagogy in schools, but can also be implemented by NGOs to getting to know the local community. Environmental citizenship can be integrated simultaneously with global citizenship if it is nurtured in the form of social learning, such as andragogy and heutagogy beyond pedagogy that focuses solely on primary and secondary education institutions.

Table 8 shows a strategic framework for streamlining NGO programs toward achieving SDGs. Each strategy has a clear and distinct output according to the contextual requirements of the SDGs. There are three strategies presented: The first is mainstreaming the SDGs in the goals of GEC and its partner institutions, the second is institutionalizing the SDGs in the capacity of GEC institutions, and the third is translating SDGs into every dimension of sustainable development.

The first strategy is to mainstream the SDGs in GEC's organizational goals as well as that of their partner institutions should consider their respective organizational capacities by matching the relevant SDGs to align with their organizational goals. However, mainstreaming SDGs by any NGO capacities needs to clearly consider the types of roles they can play, while questioning their legitimacy as to their interest in implementing this global agenda, so that the implementation of the SDGs is not confused between organizational needs and SDGs' needs. This is because the implementation of the SDGs needs to be understood as what they can do to reach the target of SDGs, whereas the needs of the SDGs need to be understood as they need to play a role in facing the transformative challenges of the SDGs.

The second strategy, institutionalizing the SDGs in the capacity of GEC institutions, takes into account the institutional capacity of a network of institutions across all sectors and organizations. By looking at the appropriate type of governance structure to be applied to a given institutional capacity, knowledge and readiness of the SDGs are able to divert all sectors' view toward their responsibility to allocate knowledge capital, social capital and transitional capital in order to make transitional institutional arrangements to shift

**Table 8** Strategic framework for streamlining NGO programs toward achieving SDGs

| Strategies | NGO program impacts | Strategic output | SDGs context |
|---|---|---|---|
| Mainstreaming the SDGs in the goals of GEC and its partner institutions | Action oriented | Integrated plan for SDGs | Increased trade-offs between SDGs |
| Institutionalizing the SDGs in the capacity of GEC institutions | Formalization of commitments | Hybrid governance | Transition from single-sector approach to cross-sector approach |
| Translating SDGs into every dimension of sustainable development | Social values | Facilitating environmental citizenship | Social inclusion, economic growth and environmental protection |

the single-sector approach toward a more inclusive cross-sector approach.

The third strategy, translating SDGs into every dimension of sustainable development, takes into account all the knowledge capital (bringing together expertise), social capital (bringing together all economic and social resources) and transitional capital (influencing the country's political decisions in determining the direction of national development) available in organizational and institutional capacities to foster environmental citizenship as a form of social value so that every decision made by an organization or institution can support a country's developmental change toward sustainable development.

In any program run by NGOs, in particular GEC itself has eight program orientations whereby the implementation of the SDGs by NGOs can be monitored only by evaluating action orientations, formalization of commitments and social values as the impacts of NGO programs. While action orientations, commitments and social values can be in many forms, these three program impacts can be used in evaluating the performance of any program undertaken by any organization or institution as a standard for achieving sustainable development.

Each strategy has output that can serve as a reference for all organizations and institutions interested in implementing the SDGs. Like the first action-oriented strategy, an integrated plan for SDGs can be proposed by combining several action plan ideas in terms of management, operations and so on that are common mechanisms for an organization to achieve common goals with stakeholders within their institutional network. For the second strategy, hybrid governance is one form of formalization of commitment to the institutional capacities of GEC as their many stakeholders comprising public–private partnerships. Hybrid governance can be formed with the mandate of a governmental sector with specific jurisdictions to call on other sectors to play a role and influence in accordance with the SDGs by establishing a special committee to monitor the implementation of the SDGs at the institutional level while making NGOs as a key driver of the SDGs implementation at the local level. The final output is the facilitation of environmental citizenship because after the formalization of commitments at all institutional and local levels, NGOs can promote the adoption of environmental citizenship as a change of attitude and behavior of all Malaysians through the education system, policy, management and action plan of all institutions in forming a harmonious unity for social inclusion, green economy for economic growth, and diversifying and prioritizing the ecosystem approach as a way to address environmental issues in Malaysia.

## 7 Conclusions

The advantages of NGOs in terms of wide range of organizational interests and aspects of society as well as overcoming other sectors are considered to be able to implement the SDGs. The existence of this third sector in development not only represents the voice of the civil society but fills the gaps of the sector and the market when the constituencies of the government and industrial sectors are limited. However, research has shown that NGOs are also a source of narcissistic attitudes and their ideological contradictions make other sectors less comfortable with the presence of NGOs as one of the major institutions of the society. At the same time, identifying the characteristics of NGOs as one of the key sectors and key institutions of the society is a major challenge as the structure and stance of the organization are too complex and changing from its point of view in development, namely its role, legitimacy, accountability, representation and a constantly evolving performance according to the current needs of institutions and the passage of time. The capacities of NGOs are also influenced by the growth in the number of branches and project management that determines the size of an NGO's operations over time as public confidence increases. The uncertainty in terms of capacities of these NGOs comes from many angles, whereby what makes it important is that the need for the SDGs is still new to the thinking of NGOs, especially in the transition from single-sector approach to cross-sector approach to increase trade-offs between the SDGs. In fact, institutionalization is also unfamiliar to the notion among institutions to the extent that it can make the implementation of the SDGs difficult to achieve at the optimal level. Therefore, streamlining NGO programs is very important to assist in the implementation of the SDGs.

As a case study, the Global Environment Centre (GEC) helped to formulate a coordination strategy for achieving the SDGs. Generally, NGOs can be tailored to three levels, namely organizational capacity, institutional capacity and sustainable development dimensions. In terms of NGOs' organizational capacities, NGOs' programs are tailored to determine the role of NGOs directly or indirectly in accordance with the relevant SDGs' targets. Each role of an NGO has its own characteristics that can be grouped into eight program orientations. The program orientations are linked to four NGOs' legitimacy assets, namely legal, pragmatic, cognitive and normative. NGOs' organizational capabilities are further strengthened by determining legitimacy environment and symbol. In terms of NGOs' institutional capacities, measuring the level of institutional partners' knowledge about the SDGs is important for NGOs to coordinate appropriate programs to increase the readiness of the SDGs among institutional partners and the local communities. Streamlining NGOs' programs through cross-sector partnership governance mechanism requires further information related to cooperation in NGOs' institutional capacities, such as forms of cooperation, levels of cooperation and benefits of cooperation, while streamlining NGOs' program through broadening social value translation mechanism requires more information related to broadening social interactions such as the context of broadening social value. From both SDGs implementation mechanisms, NGOs' programs can be streamlined through the SDGs participation domain to formalize sectoral commitments by examining the openness of their institutional partners in implementing the SDGs and selecting the most appropriate sector to inform the implementation of the SDGs. In terms of sustainable development dimensions, NGOs' programs can be streamlined by translating NGOs' program orientations into the impact of NGOs' programs.

To enhance trade-off between the SDGs, GEC can mainstream the SDGs into its organization as well as its partner organizations' goals by coproducing strategies and programs for the implementation since GEC's legitimacy environment and symbol show that these organizations have a highly interconnected and extensive constituency from local to global. In the context of shifting a single-sector approach to a cross-sector approach, it requires hybrid governance as a form of multi-sectoral governance within the institutional capability of GEC as this cross-sector partnership mechanism requires public–private and multi-sectoral cooperation. In the context of institutional capacity, GEC can enhance the institutional knowledge of SDGs by improving institutional partner's knowledge on SDGs through promotion in workplaces as well as social media. At its internal organization, GEC can promote SDGs to local communities that are directly involved in their campaign projects and provide campaign materials related to the SDGs. At its external organization, GEC can promote SDGs through CSR synergies available with its corporate partners to connect its institutional partners from different sectors. In addition to enhancing knowledge and readiness of SDGs among institutional partners, GEC also needs to get feedbacks from institutional partners on the organizational constraints and institutional capacities in implementing the SDGs, so that GEC can provide ideas of how the implementation of the SDGs can proceed according to the organizational capabilities of their respective sector.

In the context of integrating sustainable development dimensions that are social inclusion, economic growth and environmental protection, it can be translated through broadening social value mechanism to foster environmental citizenship. Social value is important in influencing social attitude and behavioral change for social adjustment since social value is an instrumental value that links organizational goals and institutions at the micro-level by enhancing social interactions related to the needs of SDGs up to the macro-level. Social adjustment can put pressure on institutions to structure institutions which is to shift a single-sector approach to a cross-sector approach. Therefore, the SDGs are a key global target today that can make social value a measurement of the sustainability performance of the UN member states.

A framework has been proposed consisting of three strategies to streamline NGO programs toward achieving SDGs, whereby the framework is intended to guide GEC or any NGOs to implement SDGs as a form of bottom-up approach by translating every requirement of the SDGs into action-oriented programs, forging hybrid governance for cooperation among GEC partner institutions and making social value the essence of fostering environmental citizenship. As NGOs have unmeasurable capacities, this strategy can give the idea for any NGO irrespective of the role and goal of the organization to streamline their programs strategically according to the needs of the SDGs. This framework is also expected to enhance the priorities of the SDGs as a measurement of project performance within an NGO and in any sector so that NGOs' project performance can be standardized despite diverse project scope.

## References


Adler, A. (1970). *Superiority and social interest: A collection of later writings*. Illinois: Northwestern University Press.

Agranoff, R., & McGuire, M. (1998). Multinetwork management: Collaboration and the hollow state in local economic policy. *Journal of Public Administration Research and Theory, 8*(1), 67–91.

Alatas, S. M. (2003). The role of NGOs and non-state actors in Malaysia's foreign policy formulation during the Mahathir era. *Akademika, 62*(1), 59–84.

Anthony, G. (1984). *The constitution of society: Outline of the theory of structuration*. Oakland: University of California Press.

Arenas, D., Lozano, J. M., & Albareda, L. (2009). The role of NGOs in CSR: Mutual perceptions among stakeholders. *Journal of Business Ethics, 88*(1), 175–197.

Arhin, A. (2016). Advancing post-2015 sustainable development goals in a changing development landscape: Challenges of NGOs in Ghana. *Development in Practice, 26*(5), 555–568.

Atack, I. (1999). Four criteria of development NGO legitimacy. *World Development, 27*(5), 855–864.

Avina, J. (1993). The evolutionary life cycles of non-governmental development organizations. *Public Administration and Development, 13*(5), 453–474.

Bäckstrand, K. (2006). Multi-stakeholder partnerships for sustainable development: Rethinking legitimacy, accountability and effectiveness. *European Environment, 16*(5), 290–306.

Banks, N., & Hulme, D. (2012). The role of NGOs and civil society in development and poverty reduction. *Brooks World Poverty Institute Working Paper*, (171).

Beardsley, M. C. (1965). Intrinsic value. *Philosophy and Phenomenological Research, 26*(1), 1–17.

Behn, R. D. (2001). *Rethinking democratic accountability*. Washington: Brookings Institution Press.

Biermann, F., Kanie, N., & Kim, R. E. (2017). Global governance by goal setting: The novel approach of the UN sustainable development goals. *Current Opinion in Environmental Sustainability, 26*, 26–31.

Boas, I., Biermann, F., & Kanie, N. (2016). Cross-sectoral strategies in global sustainability governance: Towards a nexus approach. *International Environmental Agreements: Politics, Law and Economics, 16*(3), 449–464.

Bojang, A. B. (2001). *Citizenship education: The global dimension: Guidance for Key Stages 3 and 4*. Development Education Association.

Boyer, R. (2000). *The French welfare: An institutional and historical analysis in European perspective* (No. 0007). Cepremap.

Brinkerhoff, J. M. (2003). Donor-funded government—NGO partnership for public service improvement: Cases from India and Pakistan. *Voluntas: International Journal of Voluntary and Nonprofit Organizations, 14*(1), 105–122.

Brundtland, G. H. (1987). Our common future—Call for action. *Environmental Conservation, 14*(4), 291–294.

Bryson, J. M., Crosby, B. C., & Stone, M. M. (2006). The design and implementation of Cross-Sector collaborations: Propositions from the literature. *Public Administration Review, 66*, 44–55.

Carroll, T. F. (1992). *Intermediary NGOs: The supporting link in grassroots development*. Colorado: Kumarian Press.

Chisolm, L. B. (1995). Accountability of nonprofit organizations and those who control them: The legal framework. *Nonprofit Management and Leadership, 6*(2), 141–156.

Clark, J. (1991). *Democratizing development: The role of voluntary organizations* (No. E51 C593 GTZ-PR). Colorado: Kumarian Press.

Daly, H. E. (1990). Toward some operational principles of sustainable development. *Ecological Economics, 2*(1), 1–6.

Dart, R. (2004). The legitimacy of social enterprise. *Nonprofit Management and Leadership, 14*(4), 411–424.

de Fonseka, C. (1995). Challenges and future directions for Asian NGOs. *Government-NGO Relations in Asia* (pp. 57–76). London: Palgrave Macmillan.

DiMaggio, P. J., & Powell, W. W. (1991). Introduction. The new institutionalism in organizational analysis. *The New Institutionalism in Organizational Analysis*. Chicago: University of Chicago Press, pp. 1–38.

DiMaggio, P. J., & Powell, W. W. (2004). The iron cage revisited: institutional isomorphism and collective rationality in organizational fields. *The New Economic Sociology*, 111–134.

Dobson, A. (2007). Environmental citizenship: Towards sustainable development. *Sustainable Development, 15*(5), 276–285.

Doh, J. P., & Teegen, H. (2002). Nongovernmental organizations as institutional actors in international business: Theory and implications. *International Business Review, 11*(6), 665–684.

Dowling, J., & Pfeffer, J. (1975). Organizational legitimacy: Social values and organizational behavior. *Pacific sociological review, 18* (1), 122–136.

Dunn, J. (1999). *Situating democratic political accountability. Democracy, accountability, and representation*. United Kingdom: Cambridge University Press, pp. 329–345. ISBN 0-521-64616-2.

Ebrahim, A. (2003). Accountability in practice: Mechanisms for NGOs. *World Development, 31*(5), 813–829.

Edwards, M., & Hulme, D. (1995). NGO performance and accountability in the post-cold war world. *Journal of International Development, 7*(6), 849–856.

Edwards, M., & Hulme, D. (1996). *Beyond the magic bullet: NGO performance and accountability in the post-cold war world*. Colorado: Kumarian Press.

Edwards, M. (1999). International development NGOs: Agents of foreign aid or vehicles for international cooperation? *Nonprofit and Voluntary Sector Quarterly, 28*(1_suppl), 25–37.

Espey, J., Walęcik, K., & Kühner, M. (2015). *Follow-up and Review of the SDGs: Fulfilling our commitments*. Sustainable Development Solutions Network.

Etzkowitz, H. (2008). *The triple helix: University-industry-government innovation in action*. London and New York: Routledge.

Falk, R., 1993. The Making of Global Citizenship. In: J. Brecher, et al. (Eds.), *Global visions: Beyond the new world order* (pp. 39–50). Cambridge, MA: South End Press. ISBN 0896084604.

Fisher, J. (1997). *Nongovernments: NGOs and the political development of the third world*. Colorado: Kumarian Press. ISBN-14 9781565490741.

Fowler, A. (1996). Demonstrating NGO performance: Problems and possibilities. *Development in practice, 6*(1), 58–65.

Fowler, A. (2000). Introduction beyond partnership: getting real about NGO relationships in the aid system. *IDS bulletin, 31*(3), 1–13. https://doi.org/10.1111/j.1759-5436.2000.mp31003001.x.

Frantz, T. R. (1987). The role of NGOs in the strengthening of civil society. *World Development, 15*, 121–127.

Fry, R. E. (1995). Accountability in organizational life: Problem or opportunity for nonprofits? *Nonprofit Management and Leadership, 6*(2), 181–195.

Ganesh, S. (2003). Organizational narcissism: Technology, legitimacy, and identity in an Indian NGO. *Management Communication Quarterly, 16*(4), 558–594.

Gemmill, B., Bamidele-Izu, A., Esty, D. C., & Ivanova, M. H. (2002). *Global environmental governance: Options and opportunities*. Yale Center of Forestry & Environmental Studies.

Giddings, B., Hopwood, B., & O'brien, G. (2002). Environment, economy and society: Fitting them together into sustainable development. *Sustainable Development, 10*(4), 187–196.

Glaser, G. (2012). Policy: Base sustainable development goals on science. *Nature, 491*(7422), 35.

Goldsmith, S., & Eggers, W. D. (2005). *Governing by network: The new shape of the public sector*. Washington: Brookings Institution Press.

Googins, B. K., & Rochlin, S. A. (2000). Creating the partnership society: Understanding the rhetoric and reality of cross-sectoral partnerships. *Business and Society Review, 105*(1), 127–144.

Gordenker, L., & Weiss, T. G. (1995). Nongovernmental organisations, the United Nations and global governance. Abingdon (United Kingdom) Carfax. http://www.fao.org/library/library-home/en/

Griggs, D., Stafford-Smith, M., Gaffney, O., Rockström, J., Öhman, M. C., Shyamsundar, P., et al. (2013). Policy: Sustainable development goals for people and planet. *Nature, 495*(7441), 305.

Grybaite, V., & Tvaronavičiene, M. (2008). Estimation of sustainable development: Germination on institutional level. *Journal of business economics and management, 9*(4), 327–334.

Harsh, M., Mbatia, P., & Shrum, W. (2010). Accountability and inaction: NGOs and resource lodging in development. *Development and Change, 41*(2), 253–278.

Hassan, M., Lee, K. E., & Mokhtar, M. (2019). Streamlining non-governmental organizations' programs towards achieving the sustainable development goals: A conceptual framework. *Sustainable Development, 27*(3), 401–408.

Hazlewood, P., & Bouyé, M. (2018). *Sustainable development goals: Setting a new course for people and planet*. World Resources Institute: Insights, viewed 18 September 2016, https://perma.cc/GAJ2-TGKJ.

Healey, P. (1998). Building institutional capacity through collaborative approaches to urban planning. *Environment and Planning A, 30*(9), 1531–1546.

Hezri, A., & Ghazali, R. (2011). *A fair green economy? Studies of agriculture, energy and waste initiatives in Malaysia* (No. 2). UNRISD Occasional Paper: Social Dimensions of Green Economy and Sustainable Development.

Hezri, A. A. (2016). *The sustainability shift: Refashioning Malaysia's future*. Areca Books.

Hezri, A. A., & Dovers, S. R. (2006). Sustainability indicators, policy and governance: Issues for ecological economics. *Ecological Economics, 60*(1), 86–99.

Hilhorst, D. (2003). *The real world of NGOs: Discourses, diversity and development*. London: Zed Books. ISBN-13: 978-1842771655.

Hooi, K. Y. (2013). The NGO-Government relations in Malaysia: Historical context and contemporary discourse. *Malaysian Journal of Democracy and Election Studies, 1*(1), 76–85.

Hopwood, B., Mellor, M., & O'Brien, G. (2005). Sustainable development: Mapping different approaches. *Sustainable Development, 13*(1), 38–52.

Hulme, D. (2009). The Millennium Development Goals (MDGs): A short history of the world's biggest promise.

Imperative., S. P. (2015). Social progress index 2015. *Basic Human Needs Ranking*, viewed March 12th, 2017, https://pdfs.semanticscholar.org/ddd8/1ca8944fe5757b312290505b7c8a5f6b4278.pdf.

Jayasooria, D. (2016). Sustainable development goals and social work: Opportunities and challenges for social work practice in Malaysia. *Journal of Human Rights and Social Work, 1*(1), 19–29.

Jepson, P. (2005). Governance and accountability of environmental NGOs. *Environmental Science & Policy, 8*(5), 515–524.

Kemp, R., Parto, S., & Gibson, R. B. (2005). Governance for sustainable development: Moving from theory to practice. *International Journal of Sustainable Development, 8*(1–2), 12–30.

Kenis, P., & Provan, K. G. (2009). Towards an exogenous theory of public network performance. *Public Administration, 87*(3), 440–456.

Kickert, W. J., Klijn, E. H., & Koppenjan, J. F. (1997). *Managing complex networks: Strategies for the public sector*. New York: Sage.

Klasen, S., & Fleurbaey, M. (2018). Leaving no one behind: Some conceptual and empirical issues. UN Department of Economic and Social Affairs.

Kolk, A., Van Tulder, R., & Kostwinder, E. (2008). Business and partnerships for development. *European Management Journal, 26*(4), 262–273.

Korten, D. C. (1984). Strategic organization for people-centered development. *Public Administration Review, 44*(4), 341–352.

Korten, D. C. (1990). *Getting to the 21st century: Voluntary action and the global agenda*. Colorado: Kumarian Press.

Koschmann, M. A., Kuhn, T. R., & Pfarrer, M. D. (2012). A communicative framework of value in cross-sector partnerships. *Academy of Management Review, 37*(3), 332–354.

Kumar, S., Kumar, N., & Vivekadhish, S. (2016). Millennium development goals (MDGS) to sustainable development goals (SDGS): Addressing unfinished agenda and strengthening sustainable development and partnership. *Indian Journal of Community Medicine: Official Publication of Indian Association of Preventive & Social Medicine, 41*(1), 1.

Lele, S. M. (1991). Sustainable development: A critical review. *World Development, 19*(6), 607–621.

Lewis, D. (1998). Development NGOs and the challenge of partnership: Changing relations between North and South. *Social Policy & Administration, 32*(5), 501–512.

Lewis, D. (2010). Disciplined activists, unruly brokers? Exploring the boundaries between non-governmental organizations (ngos), donors, and the state in Bangladesh. *Varieties of Activist Experience: Civil society in South Asia*, 159–180.

Lewis, D., & Kanji, N. (2009). *Non-governmental organizations and development*. United Kingdom: Routledge.

Li, B., & Akintoye, A. (2003). *An overview of public-private partnership. Public Private Partnerships: Managing Risks and Opportunities* (pp. 3–30). United State: Blackwell Publishing.

Lindenberg, S. (2001). Intrinsic motivation in a new light. *Kyklos, 54*(2–3), 317–342.

Lister, S. (2003). NGO legitimacy: Technical issue or social construct? *Critique of Anthropology, 23*(2), 175–192.

Lu, Y., Nakicenovic, N., Visbeck, M., & Stevance, A. S. (2015). Policy: Five priorities for the UN sustainable development goals. *Nature News, 520*(7548), 432.

Maldonado, V. (2010). Achieving the MDGs through quadruple helix partnerships: University—government—industry—third sector collaboration. Global University Network for Innovation, viewed 15 May 2017, http://www.guninetwork.org/articles/achieving-mdgs-through-quadruple-helixpartnerships-university-government-industry-third.

Maldonado, V., Lobera, J., & Escrigas, C. (2009). The role of higher education in a new quadruple helix context. *Triple Helix, 7*, 17–19.

Mandell, M. P. (2001). *Getting results through collaboration: Networks and network structures for public policy and management*. Santa Barbara: ABC-CLIO.

McClintock, C. G., & Allison, S. T. (1989). Social value orientation and helping behavior 1. *Journal of Applied Social Psychology, 19*(4), 353–362.

Mebratu, D. (1998). Sustainability and sustainable development: Historical and conceptual review. *Environmental Impact Assessment Review, 18*(6), 493–520.

Messick, D. M., & McClintock, C. G. (1968). Motivational bases of choice in experimental games. *Journal of Experimental Social Psychology, 4*(1), 1–25.

Najam, A. (1996). NGO accountability: A conceptual framework. *Development Policy Review, 14*(4), 339–354.

Nanthagopan, Y., Williams, N. L., & Page, S. (2016). Understanding the nature of project management capacity in Sri Lankan non-governmental organisations (NGOs): A resource-based perspective. *International Journal of Project Management, 34*(8), 1608–1624.

Ocampo, J. A. (2015). A post-2015 monitoring and accountability framework. *Department of Economic & Social Affairs, CDP Background Paper*, 27, 1–17.

Olson, M. (1971). *The logic of collective action: Public goods and the theory of groups*. Cambridge, MA: Harvard University Press. ISBN 0674537513.

Oxfam. (1997). A curriculum for global citizenship: Oxfam's development education program, viewed 22 June 2017 https://unesdoc.unesco.org/ark:/48223/pf0000197671.

Parish, F., Sirin, A., Charman, D., Joosten, H., Minaeva, T., Silvius, M., et al. (2008). *Assessment on Peatlands, Biodiversity and Climate Change—Main Report*. Global Environment Centre, Kuala Lumpur & Wetlands International, Wageningen, viewed 10 Mac 2017 http://www.peat-portal.net/index.cfm?&menuid=125.

Przeworski, A., Stokes, S.C., & Manin, B. (1999). *Democracy, accountability, and representation* (Vol. 2). United Kingdom: Cambridge University Press.

Putnam, H. (2002). *The collapse of the fact/value dichotomy and other essays*. Cambridge MA: Harvard University Press.

Ramstad, Y. (1989). Reasonable value versus instrumental value: Competing Paradigms in Institutional Economics. *Journal of Economic Issues, 23*(3), 761–777.

Redclift, M. (2005). Sustainable development (1987–2005): An oxymoron comes of age. *Sustainable Development, 13*(4), 212–227.

Reimann, K. D. (2006). A view from the top: International politics, norms and the worldwide growth of NGOs. *International Studies Quarterly, 50*(1), 45–67.

Rein, M., & Stott, L. (2009). Working together: Critical perspectives on six cross-sector partnerships in Southern Africa. *Journal of Business Ethics, 90*(1), 79–89.

Rethemeyer, R. K. (2005). Conceptualizing and measuring collaborative networks. *Public Administration Review, 65*(1), 117–121.

Robert, K. W., Parris, T. M., & Leiserowitz, A. A. (2005). What is sustainable development? Goals, indicators, values, and practice. *Environment: Science and Policy for Sustainable Development, 47* (3), 8–21.

Sachs, J. D. (2012). From millennium development goals to sustainable development goals. *The Lancet, 379*(9832), 2206–2211.

Schwartz, F. J., & Pharr, S. J. (2003). *The state of civil society in Japan*. United Kingdom: Cambridge University Press.

Seitanidi, M. M., & Crane, A. (2009). Implementing CSR through partnerships: Understanding the selection, design and institutionalisation of nonprofit-business partnerships. *Journal of Business Ethics, 85*(2), 413–429.

Selsky, J. W., & Parker, B. (2005). Cross-sector partnerships to address social issues: Challenges to theory and practice. *Journal of Management, 31*(6), 849–873.

Sjöstrand, S. E. (1992). On the rationale behind "irrational" institutions. *Journal of Economic Issues, 26*(4), 1007–1040.

Smillie, I. (1994). Changing partners: Northern NGOs, northern governments. *Voluntas: International Journal of Voluntary and Nonprofit Organizations, 5*(2), 155–192.

Sogge, D., Biekart, K., & Saxby, J. (1996). *Compassion and calculation: The business of private foreign aid*. London: Pluto Press.

Soh, M. C., & Tumin, M. (2017). The history and development of non-governmental organization (NGO) health in Malaysia for the year 2015. *HISTORY: Journal of the Department of History, 26*(2).

Sørensen, E., & Torfing, J. (2007). Theoretical approaches to democratic network governance. *Theories of democratic network governance* (pp. 233–246). London: Palgrave Macmillan.

Spijkers, O., & Honniball, A. (2014). MDGs and SDGs: Lessons Learnt from global public participation in the drafting of the UN development goals. *Vereinte Nationen: German Review on the United Nations, 62*(6), 251–256.

Spitz, G., Kamphof, R., & Van Ewijk, E. (2015). Wait-and-see or take the lead. *Approaches of Dutch CSOs to the Sustainable Development Goals*.

Stiglitz, J. E., Sen, A., & Fitoussi, J. P. (2017). Report by the commission on the measurement of economic performance and social progress.

Suchman, M. C. (1995). Managing legitimacy: Strategic and institutional approaches. *Academy of Management Review, 20*(3), 571–610.

Tarlock, A. D. (2001). Ideas Without Institutions: The Paradox of Sustainable Development. *Indiana Journal Global Legal Studies, 9*, 35.

Teegen, H., Doh, J. P., & Vachani, S. (2004). The importance of nongovernmental organizations (NGOs) in global governance and value creation: An international business research agenda. *Journal of International Business Studies, 35*(6), 463–483.

Tool, M. R. (1993). The theory of instrumental value: Extensions, clarifications. *Institutional economics: Theory, method, policy* (pp. 119–172). Dordrecht: Springer.

Turner, S. (1998). Global civil society, anarchy and governance: Assessing an emerging paradigm. *Journal of Peace Research, 35* (1), 25–42.

TWI2050. (2018). *Transformations to achieve the sustainable development goals. Report prepared by e World in 2050 initiative*. International Institute for Applied Systems Analysis (IIASA), Laxenburg, Austria, viewed 10 January 2019, http://www.iiasa.ac.at/web/home/research/twi/TWI2050.html.

UNDP. (2014). Delivering the post—2015 development agenda: Opportunities at the national and local levels, viewed 25 April 2016, http://www.undp.org/content/dam/undp/library/MDG/Post2015/UNDP-MDG-Delivering-Post-2015-Report-2014.pdf.

United Nations. (2015). The Millennium Development Goals Report 2015. ISBN 978-92-1-101320-7.

Vandemoortele, J. (2011). The MDG story: Intention denied. *Development and Change, 42*(1), 1–21.

Vakil, A. C. (1997). Confronting the classification problem: Toward a taxonomy of NGOs. *World Development, 25*(12), 2057–2070.

van Waart, P., Mulder, I., & de Bont, C. (2016). A participatory approach for envisioning a smart city. *Social Science Computer Review, 34*(6), 708–723.

van Zeijl-Rozema, A., Cörvers, R., Kemp, R., & Martens, P. (2008). Governance for sustainable development: A framework. *Sustainable Development, 16*(6), 410–421.

Weber, E. P. (1999). The question of accountability in historical perspective: From Jackson to contemporary grassroots ecosystem management. *Administration & Society, 31*(4), 451–494.

Weiss, M. L., & Hassan, S. (2003). Introduction: From moral communities to NGOs. *Social movements in Malaysia* (pp. 12–27). United Kingdom: Routledge.

Weitz, N., Nilsson, M., & Davis, M. (2014). A nexus approach to the post-2015 agenda: Formulating integrated water, energy, and food SDGs. *SAIS Review of International Affairs, 34*(2), 37–50.

Yap, N. (1990). NGOs and sustainable development. *International Journal, 45*(1), 75–105.

Young, O. R., Underdal, A., Kanie, N., & Kim, R. E. (2017). *Goal setting in the Anthropocene: The ultimate challenge of planetary stewardship* (p. 53). Governing Through Goals: Sustainable Development Goals as Governance Innovation.

Zainon, S., Atan, R., & Bee Wah, Y. (2014). An empirical study on the determinants of information disclosure of Malaysian non-profit organizations. *Asian Review of Accounting, 22*(1), 35–55.

Printed in the United States
By Bookmasters